Women at Imperial College

Past, Present and Future

Women at Imperial College

Past, Present and Future

To Fred Felix Editor
2018/2019

Anne Barrett
Imperial College London, UK

Anne Barrett

World Scientific
NEW JERSEY • LONDON • SINGAPORE • BEIJING • SHANGHAI • HONG KONG • TAIPEI • CHENNAI • TOKYO

Published by

World Scientific Publishing Europe Ltd.

57 Shelton Street, Covent Garden, London WC2H 9HE

Head office: 5 Toh Tuck Link, Singapore 596224

USA office: 27 Warren Street, Suite 401-402, Hackensack, NJ 07601

Library of Congress Cataloging-in-Publication Data
Names: Barrett, Anne (Archivist) author.
Title: Women at Imperial College : past, present and future / by Anne Barrett
(Imperial College London, UK).
Description: [Hackensack] New Jersey : World Scientific, [2017] | Includes index.
Identifiers: LCCN 2016041699 | ISBN 9781786342621 (hc : alk. paper)
Subjects: LCSH: Women scientists--Great Britain--Biography. | Women in science--Great Britain. |
Women engineers--Great Britain--Biography. | Women in engineering--Great Britain. |
Imperial College, London--Biography.
Classification: LCC Q141 .B294 2017 | DDC 509.2/5209421--dc23
LC record available at https://lccn.loc.gov/2016041699

British Library Cataloguing-in-Publication Data
A catalogue record for this book is available from the British Library.

Desk Editors: Kalpana Bharanikumar/Mary Simpson

Typeset by Stallion Press
Email: enquiries@stallionpress.com

Printed in Singapore

Author Biography

Anne Barrett MA, AIC Imperial College Archivist and Corporate Records Manager, has extensive experience of working with scientific archives. At Imperial College, this expertise has been invaluable in satisfying the diverse research requirements of those seeking to access sometimes obscure institutional resources. Anne's role includes the advancement of Imperial's achievements through the provision of biographical data and the promotion of the scientific discoveries of its alumni; the message being disseminated online, in books, invitations to speak at international conferences and join relevant committees, and is showcased in exhibitions.

Externally, she has national and international roles in archival standards, and archives and history advisory bodies. She is Chairman of Archives for London, an advocacy body for archives, and Chairman of Trustees Centre for Scientific Archives, a science, technology, engineering and medicine heritage guardian.

Acknowledgements

My grateful thanks to all who have read and approved their biographies; for permissions to: The Oxford Dictionary of National Biography (ODNB) for enabling me to quote from their biographical texts with full acknowledgement; The Royal Society; The Pathological Society of Great Britain; The British Geological Survey; The *Voices of Science*: Oral History of British Science Programme, Director Rob Perks, and The *National Life Stories* Programme; The British Library; Oxford Brookes University Video Archives; Girton College Cambridge; Sydney Sussex College Cambridge; Geoff and Marelene Rayner-Canham, Bill Brock. Thanks also to supporters Graham Rainford, Philippa Marks, Curator, Bookbindings British Library for advice and assistance, Mrs Angela Darwin, my team of Assistant Archivists at Imperial, particularly Catherine Harpham, and also to past members, Eilís McCarthy, Bryony Hooper, Ruth Kusionwicz, and to Jon Hancock, Head of Central Secretariat. To Dame Julia Stretton Higgins, Professor Dot (Dorothy) Griffiths OBE and Professor Alice Gast, Hailey Smith, Abby Bolt (cover design); The London Library for allowing their books to be on loan for long periods and for their e-library. To World Scientific and particularly editors Mary Simpson and Kalpana Bharanikumar. Every effort has been made to trace copyright ownership, should anyone wish to assert a right, please contact the author.

Contents

Foreword

As this book goes to press in 2016, Imperial College is now in the happy position of having too many women to be able to include them all! The biographies which are included, range from the 19th to the 21st centuries, and reveal career paths and progress in development of women's roles at Imperial. Whilst it is a pity that more cannot be recognised individually, the steadily increasing numbers and influence of women are celebrated each year collectively during **Women@Imperial week**, which includes International Women's Day each March.

Initiatives to attract, reward and retain women staff are now embedded in Imperial's working life, driven by the high level "Academic Gender Strategy Committee". This book, though necessarily with a limit to its scope, adds to the many ways that the College recognises women and their continued contribution to the University and to its mission.

A University such as Imperial, functions through the collaborative effort of everyone working together, so while celebrating women at Imperial is important now, in future the fewer distinctions that are made between members, the better it will be for the mission of the College. The next generations will take up this dynamic view and continue developing Imperial as a good place to be for all its members.

Professor Dame Julia Stretton Higgins
July 2016

The Provost's Envoy for Gender Equality

This part time post was created in January 2014. The first post holder was Professor OBE Dot Griffiths who took the role after she stepped down as Dean of the Business School. The creation of this post signalled a step change in the College's commitment to gender equality. The first significant development had been, in 1998, the creation of the Academic Opportunities Committee chaired, initially, by Professor Dame Julia Stretton Higgins (who did so much to create the hugely influential Athena SWAN scheme). The second was the funding of a full time post in HR to support College and Department SWAN submissions. The role of the Provost's Envoy was the third. It created a senior, visible and accountable position to challenge the continued over-representation of men.

Importantly, the post reported to the Provost and worked through the Academic Gender Strategy Committee which was chaired by the Provost and on which all the Faculty Deans sat. It meant that the role had serious senior College support which could deliver traction on actions and policies. At this time, the College had both a very supportive Provost (James Stirling) and President (Alice Gast).

In the first three years, Dot, working with others, enabled or supported the creation of a raft of activities and policies, for example: Women@ Imperial Week (2015 onwards, a week long celebration of Imperial women held annually as the week of International Women's Day), focus groups and questionnaires to all staff about support for women which led to a Gender Summit (2016), the renewal of the College Diversity Pledge (2016), changes to recruitment to encourage more female applications and other policy changes (2016), funding for carer's costs at conferences (2016).

The role has acted as a focal point for gender related activities in the college. This not only has given them greater visibility and traction but also a central point of reference.

Professor Dot Griffths OBE
Provost's Envoy for Gender Equality
July 2016

Introduction

Introduction to Women@Imperial: The Imperial College Story

Contrary to popular opinion, there have been women working and studying at Imperial since the 19th century. The biographies and discussions which follow will dispel the myth of there never having been women at Imperial. They will show that they are very much a significant part of Imperial today, and for the future!

Women have always been admitted to Imperial, in many roles: as relatives assisting the males in their scientific work, in their own right as students, research assistants, laboratory assistants, technical artists and lecturers and professors. Besides these are a whole range of careers that administer and support the scientific roles. The paths taken to attain these official positions may not have been direct, but this only reveals that there are many routes to employment associated with science, and indeed, many types of career. It is not necessary to remain in the mainstream or academia to follow such a profession. The field may embrace a wide variety of occupations ranging from administration, Imperial College Head Quarters and governance to compliance, legal and research pursuits, human resources, departmental work, student support, and as photographers, in estates roles, in occupational health, safety, libraries, ICT, and archives and records management.

Imperial College Chronology

To set the scene: Imperial College is now, 110 years old (2017). It was founded from four original constituent Colleges: The Royal College of Chemistry, (RCC) (1845), The Royal School of Mines, (RSM) (1851), subsuming the RCC; The Royal College of Science, (RCS) (1881) and The City and Guilds College, (C&GC) (1884), becoming Imperial College in 1907.[i] Imperial is a legacy of the vision of Prince Albert to create a cultural quarter of science and the arts in South Kensington, funded by the huge profits of the Great Exhibition of 1851. Imperial College expanded in the later 20th century to include medical schools: St Mary's Hospital Medical School (SMHMS) joined in 1988, then other west London schools, Charing Cross and Westminster Hospital Medical School (CXWMS), National Heart and Lung Institute (NHLI) and Brompton Hospital, The Royal Post Graduate Medical School (RPMS) and Hammersmith Hospital, to form the Imperial College School of Medicine in 2000, and also Wye College, which subsumed Swanley Horticultural College in 1945. These institutions will be referred to throughout the text.

Women and the Royal College of Chemistry

The RCC was founded by subscription. Some subscribers were women, and were named separately from their husbands, such as Lady Davy (wife of Sir Humphrey Davy, Chemist) and some alone, such as Mrs Orleton Cludde of Wellington Salop. These were life members as they gave a donation of 20 guineas. (Today, a guinea equals £1 and 1 shilling = 5 pence)

Following on from subscribers,[ii] other women also assisted its development, initially as family members of the first Professor of Chemistry, August Wilhelm von Hofmann. He had married into the German chemical family of Liebig. The first of his four wives was Helen Moldenhauer, the niece of Liebig's wife. Hofmann and Helen married in 1846, a year after his appointment to the RCC. Hofmann's sister Hannah, later joined him at the RCC, undertaking some assaying work as well as domestic assistance.

Women and the Royal School of Mines

The wives of the RSM lecturers assisted with their researches and accompanied them on field trips. There were complaints from colleagues on a field trip about Edward Forbes[iii] and his wife, who shortly after their marriage, annoyingly to their colleagues, joined the daily surveying trips somewhat late in the mornings. T.H. Huxley[iv] took his wife sea dredging on their honeymoon! Women visited the Museum of Practical Geology attached to the RSM to view the palaeontological objects, and also attended lectures.

Women Attending Lectures

Occasional Lectures

In the 19th century, several women attended classes as occasional students, albeit often with their fathers or husbands in the first instance. The subjects were taught in the same way as to matriculated students, (or registered students) therefore, the lectures were by no means 'popular' or easy to understand.

From 1851 when the RSM opened, women took the courses for Occasional Students. The first two women attending natural history lectures being a Miss Wollaston and a Miss Annabella Homeria Pollock who was accompanied by her widowed father, Field Marshal Sir George Pollock. The interesting story here is that Miss Wollaston became the second Lady Pollock in the following year when the registers show they all studied palaeontology. Presumably, the three had attended the first years lectures together, divided only in name. A Mrs Wilkinson — with no known male companion — attended geology classes in 1854–1855. There are records of several unmarried women attending other courses alone too, including Miss Garratt who became Elizabeth Garratt Anderson, campaigner for medical education for women. She took natural history and physics 1861–1862. Several of the women undertook courses over some years, which provided a good grounding in subjects such as physics, as in the case of Lady Maria Forester who attended classes between 1862 and 1864.

Matriculated Students

Later in the century, women attended as matriculated students. The first Associate of the Royal College of Science (ARCS) was a chemistry and biology student, Elizabeth Healey, who later in life was awarded the MBE. In 1893, Marion Whiteford Acworth née Stevenson, RCS was the first woman to gain an ARCS in Physics. In that year, she married Joseph Acworth, a Chemist who had also attended the RCS. His interest was in photographic processes, particularly the dry plate method, leading to him founding the Imperial Dry Plate Company. (No relation to the Imperial College name.) The Acworth's researched and wrote papers together. Marion pushed scientific boundaries by publishing anonymously, *The Great Delusion: A Study of Aircraft in Peace and War* by Neon, in 1927. She argued that in the new world of aviation, it was not necessary for there to be a military fleet, as civilian aircraft could be adapted for use in any conflict.

1st Women Graduates

One of the first students at C&GC (opened to students in 1885), was a Miss A.G. Heath who took chemistry 1885–1887 and gained the Associateship of the City and Guilds Institute (ACGI) in 1887. Subsequently she taught at the North London Collegiate School. And so it continues as Imperial gains in reputation as a welcoming place for women, more join Imperial as both students and staff. This includes those from overseas, some as refugees from conflict and untoward political situations, for example, before and during WWII.

Women students such as Mariam Oommen, who lectured in Chemistry at Madras University India (the Women's Christian College), attended the RCS 1927–1930. She took RCS chemistry courses, gaining a PhD and DIC, then returned to India to work as a lecturer. Whilst at the RCS Mariam worked closely with Martha Annie Whiteley, the notable chemist and mentor, as did many other women, some of whom have separate entries in this book.

Other students went on to become school mistresses, such as Ida Clare McMechan, RCS 1924–1926. Ida became Senior Science Mistress at The Queen Mary High School, Liverpool. Others became headmistresses or Principals, such as Marjorie Winifred Odell, RCS 1925–1928, Principal Mabel Fletcher Technical College, Liverpool. Others are discussed later.

Scientific Education for School Teachers

The Science Teacher in training scholarships provided by the Department of Science and Art[v] in the 1890s, enabled school science teachers to attend the constituent colleges in South Kensington. This system was instituted to raise the standard of science teaching throughout the country. They allowed '*country teachers to gain experience of the latest discoveries and newest experiments. They attend on the understanding that they do their work thoroughly. They are required to take notes and afterwards write a full report of the lecture and take part in practical work. Their reports are examined and marked by the lecturers .... The main purpose of these courses is to enable teachers to give practical instruction the sciences.*'[vi] One such scholarship place was awarded in 1873 to Mary Anne Gahan from the Roscrea YMCA Institution County Tipperary, Ireland.[vii]

Academic Women at Imperial

As we have seen, women had followed academic careers at College from the late 19th century. The first woman to hold a teaching role at Imperial was Martha Annie Whiteley, OBE, who became an Assistant Professor, (or a Reader in today's terminology). The first woman Professor to be appointed was Helen Kemp Porter in 1959, this was to the Chair of Plant Physiology, the second was Joan Woodward, appointed Professor of Industrial Sociology in 1970 and the third, Janet Vida Watson to a Personal Chair in Geology in 1974 and the fourth, Anita Bailey to the Kodak Chair of Interface Science in 1976. (See separate entries for all these named.) Others included the redoubtable Miss Letita Chitty who worked in the Civil Engineering Department for 28 years. She has the claim of being one of the few women scientists to be interviewed by *Vogue* for an article in 1969.[viii] She was also the first woman to be appointed to the Institution of Civil Engineers Technical Committees in 1958.

Imperial provided two principals of Swanley Horticultural College in the early 20th century, Frances Micklethwait and Kate Barratt. (See separate entries for named women.)

There are many more ‘firsts’ among Imperial College women, in becoming Fellows of learned societies and in many other achievements in their careers and lives, which will be discussed in their biographies.

> That Imperial is a good place to be is shown by the number of staff who remain for long periods, or their whole careers. This is because there is the flexibility and the opportunity to work collaboratively and creatively with colleagues and students at Imperial, and also internationally. Helen ApSimon CBE is Professor of Air Pollution Studies in the Centre for Environmental Policy and has worked at Imperial, some of the time at Silwood Park, for over 30 years. Just before her Inaugural Lecture in 2003, Helen said she was ‘honoured to have been appointed to this rank at Imperial College, and thankful to the wide range of people with whom I have worked with from very different scientific disciplines, who have helped me to achieve it.’[ix]

An example of a very long serving administrative staff member, retiring only in 2016, is Linda Jones MBE AIC, who joined the Physics Department in 1968 as a secretary — and left as Physics Operations Manager.

Linda playing in the Physics Department Secretaries vs. Postgraduates football match c1970s.[x]

"It's been a blast", she said at her retirement event. "The Department is large but wonderfully collegiate — and remains so. I am very proud to have been a part of it."[xi]

Women who took degrees or spent part of their careers at Imperial, have left for jobs in varied fields of science and engineering, and for other roles, in diverse organisations. This is a small selection of roles and organisations:

Chief Medical Officer for England, first woman post holder, Chief Scientific Adviser to the Department of Health, Dame Sally Davies FRS[xii]; The first woman to fly with the Red Arrows Aerobatic Display Team in 2010, Kirsty Moore (MSc Aeronautics 1999), and others have joined the RAF in roles such as flight engineers; The Health and Safety Executive; the Science Museum and Atomic Energy Agency, including roles as technical and library staff.

As leaders in the engineering training organisations: Judith Walker Materials MSc (1969–1970) and first Woman President Imperial College Student Union 1970–1971, has worked for the Engineering Judith Walker.

Council/Engineering UK where she was "recognised as a champion of technicians and IEngs and as knowledgeable on education and training of engineers." AS CEO of the Institution of Highway Engineers, for over 30 years to 2013, Judith was involved with "all aspects of leading, developing and sustaining a small but active professional engineering institution, not just the strategy, but the hands on admin as well."[xiii]

As journalists for national newspapers, in Rebekah Higgett's case, combining a blog in *The Guardian* newspaper, the H Word, with a

Lectureship in the History of Science at Kent University. An Imperial woman television presenter is Maggie Aderin Pocock MBE, a copresenter of the BBC's *The Sky at Night* (BSc Physics 1987–1990, Mechanical Engineering PhD 1995).

Imperial women are also Parliamentarians, as UK Members of Parliament: Karen Bradley[xiv] (BSc Mathematics) was appointed to the role of Secretary of State for Culture Media and Sport in the new Conservative Government of July 2016, Chi Onwurah[xv] (MSc Electrical Engineering) MP for Newcastle since 2010 and Shadow Minister for Culture Media and Sport 2015–, and Dame Joan Ruddock[xvi] (BSc Chemistry and Botany) MP for Lewisham London 1987–2015, first Minister for Women in a Labour Government 1997. There was also a Member of the European Parliament, for Portugal 2009–2014, Maria da Graça Carvalho (PhD Energy, Environment and Sustainable Development 1983).

Institutional and industrial governance is another sector women associated with Imperial are very much involved with. Dame Mary Archer, Chemistry PhD 1968, exemplifies this connection: Science Museum Group Chairman of the Board of Trustees 2015–; Imperial College Health Partners Chairman Expert Advisory Board 2013–; President National Energy Foundation 1999– Solar Energy Society 1999–; Guild of Church Musicians 1989–. Dame Margaret Weston[xvii] combined governance of Imperial with her roles at the Science Museum London. A chartered electrical engineer, she was a Governor of Imperial College, 1974–1990, (awarded a Fellowship of Imperial College in 1975,) whose career at the Science Museum, from 1955, culminated in her becoming its Director 1973–1986.

They also work in other areas, such as the financial sector and industry: Mais Najib (Executive MBA 2008) works in health economics and market access[xviii]; Rima Sen Mortemart (Chemistry with Management 1993) is a consultant, analysing non-financial assets and long term contingency planning for multinational corporations.[xix]

Catherine Liao (Executive MBA 2013) works for a large IT company, in Marketing but also demonstrates her entrepreneurial skills, creating a mobile wine business startup before selling that and moving into marketing, influenced by her course at Imperial. Another example of entrepreneurship is with Farah Naz (Biochemistry 1999), in the cosmetics industry, as she founded ECX1 Cosmetics.[xx] These are just some of the areas past and current Imperial women have succeeded in and will continue to succeed in. As new global opportunities occur, they will no doubt be found in those areas too.

This book celebrates the women, their work and Imperial College's evolving affirmation for Women@Imperial.

Endnotes

i A Pictorial History of Imperial College Anne Barrett and R.G. Williams 1988.

ii Royal College of Chemistry List of Subscribers Archives Imperial College.

iii Edward Forbes (1815–1854) Palaeontologist, Naturalist, Biogeographer.

iv Thomas Henry Huxley (1825–1895) Biologist, educationalist, coined the term Agnostic, known as 'Darwin's Bulldog' for defending Charles Darwin's theory of evolution against critics.

v The 1st British Government body responsible for the administration and funding of art and science policy and institutions.

vi Bernard H. Becker Scientific London 1874.

vii Mary Anne's brother William P. Gahan Royal Exhibition to RSM 1872, became a teacher after this. From The Scientific Lady, Chapter: Science Schools for Girls of the Industrial Class.

viii Vogue October 15th, 1969.

ix Imperial College Press Release Judith H. Moore 28 January 2003. Retrieved from http://www.imperial.ac.uk/college.asp?P=3903.

x Retrieved from http://www.imperial.ac.uk/centenary/memories/jones.shtml.

xi Retrieved from http://www3.imperial.ac.uk/newsandeventspggrp/imperialcollege/naturalsciences/newssummary/news_21-4-2016-15-49-22.

xii Dame Sally Davies (1949–) DBE FRS. 2014 Fellow Royal Society 2011 Emeritus Professor Imperial College 2010 Chief Medical officer of Health; 2010 voted 6th most powerful woman in the UK BBC Woman's Hour; 2009 Dame Commander British Empire 2009 Fellow Faculty of Medicine Imperial College; 2006 Established the National Institute for Health Research 1997 Professor of Haemoglobinopathies 1985 haematologist Central Middlesex Imperial College School of Medicine.

xiii Judith Walker Profile Websearch 20150127. Retrieved from https://www.linkedin.com/in/judith-walker-2618a040.

xiv Karen Bradley (1970–) BSc; 2010– MP Conservative Party; 2016– Secretary of State for Culture, Media and Sport; 2014–2016 Parliamentary Under

Secretary of State in the Home Office; 2013–2014 Government Whip (Junior Lord Commissioner of the Treasury), 2012–2013 Assistant Government Whip; MP (Conservative Party) for Staffordshire Moorlands 2010–; Career: Financial Sector, Tax Consultant; BSc Mathematics Imperial College 1991.

xv Chi Onwurah (1965–) BSc, MBA; MP Labour Party 2010–, 2010–2013 Shadow Minister Business, Innovation and Skills 2015–; 2013– Shadow Cabinet Office Minister for Digital Government, Cyber Security and Social Enterprise; 2014 commissioned Review of Digital Government; Shadow Minister Cabinet Office 2103–2015; Shadow Minister Culture Media and Sport 2015–; Cochair, Parliamentary Internet, Communications and Technology Forum; Board member, Parliamentary Office of Science and Technology; Member International Advisory Board Open University Business School (OUBS); Member Chatham House. As Shadow Minister for Business, Innovation and Skills concerned with women in science and STEM initiatives; Career: Electrical Engineer telecoms, including business and developing markets; 1983–1987 MSc Electrical Engineering Imperial College.

xvi Dame Joan Ruddock (1943–) DBE, PC, BSc, 1987–2015 MP Labour Party; 2012 Dame Commander British Empire; 2010 Privy Councillor; 1987–2015 MP for Lewisham London; 1997 1st Minister for Women in a Labour Government, assisted creation of the Equalities Unit, to improve childcare, develop action against violence to women; Honorary Fellowships: Goldsmiths University of London; 2014 Trinity Laban Conservatoire of Music and Dance for services to Laban's Board and fundraising; BSc Chemistry & Botany Imperial College.

xvii Dame Margaret Kate Weston (1926–) DBE FMA 1973–1986 Director Science Museum London; Roles as Assistant Keeper, Deputy Keeper to Keeper Science Museum from 1955. Created The National Media Museum Bradford whilst Director.

xviii Retrieved from https://www.imperial.ac.uk/alumni/imperial-alumni/alumni-stories/mais-najib/.

xix Retrieved from https://www.imperial.ac.uk/alumni/imperial-alumni/alumni-stories/rima-sen-mortemart/.

xx Retrieved from https://www.imperial.ac.uk/alumni/imperial-alumni/alumni-stories/farah-naz/.

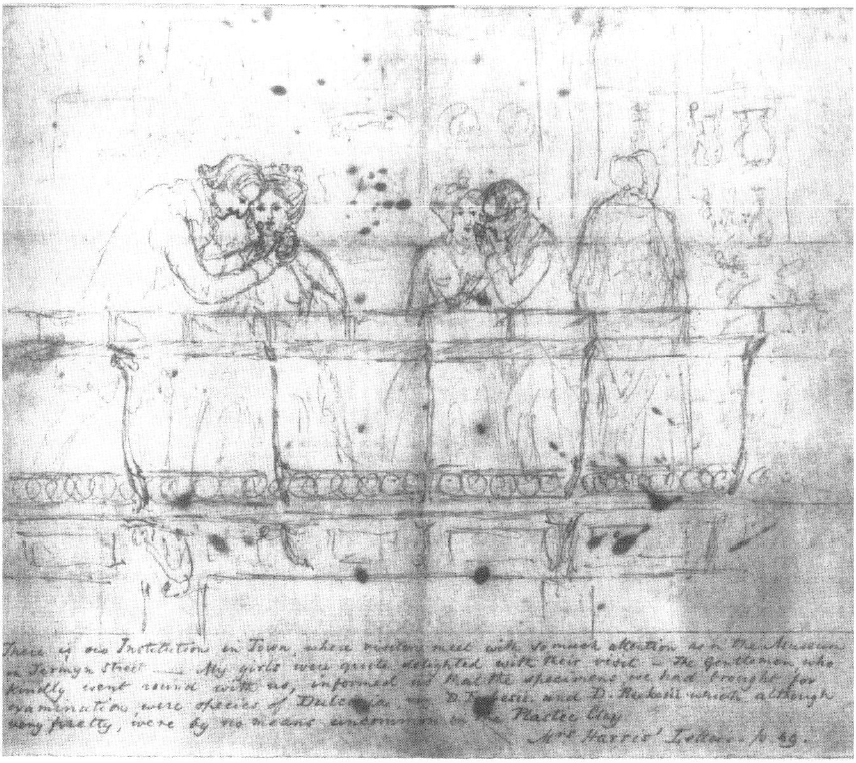

Museum of Practical Geology, Royal School of Mines. Professor Edward Forbes Palaeontologist and Trenham Reeks Registrar, describing exhibits. Amusing caption by Forbes calling himself Forbsii and Trenham Reeksii as though they were specimens themselves. Drawing by Forbes

Cartoon T.H. Huxley and wife Henrietta inland sea dredging on their honeymoon drawn by Huxley

Biographies

Geology

Lady Rachel MacRobert Workman

Name and Post Nominals

Rachel Workman Lady MacRobert (1884–1954) FGS, FMinS, FRSA, BSc.

First Imperial woman to be elected to the Fellowship of the Geological Society 1919.[i]

Specialism

Igneous Petrology, Mineralogy, Glacial Geomorphology.

Education

Privately Germany; Cheltenham Ladies College; Royal Holloway College for Women University of London; Edinburgh University; Mineralogical Institute Oslo; Royal School of Mines 1909–1912.

Career Awards

Life Fellow Geological Society, 1938.

About Her

Born in America, daughter of Dr William Hunter Workman (1847–1937), Himalayan Explorer, and Fanny Bullock Workman (1859–1925), a climber, recognised by the Royal Geographical Society. The family travelled widely meaning that Rachel was educated in Germany and Britain. Her interests in geology came from her travels. She studied the subject seriously at Royal Holloway College, Edinburgh University where she took geology and economics and the Royal School of Mines (RSM). Here in 1910, she caused consternation by apparently being the only woman ever to attend mining lectures at the RSM![ii]

In 1911, Rachel married Sir Alexander MacRobert (1859–1925) of Douneside and Cromar, Scotland. He had interests in Indian mills and founded the British India Corporation, of which she became a Director. She maintained her geological work after her marriage and having her three children. Sadly Rachel lost all three sons to flying. Her response to them being killed in flying accidents, two during WWII, was to provide a bomber and four fighter aircraft to the RAF, known as the MacRobert' s Reply. She also allowed the RAF to use her mansion in Scotland as a hospital and rest centre during the war. Rachel became a noted cattle breeder and agriculturalist postwar. The MacRobert Trust has several interests, including continuing contact with the RAF.[iii]

Endnotes

[i] Eight women were elected in 1919, some as subscription fillers Rachel Workman MacRobert claimed.

[ii] Rachel Workman letter to her Fiancé Sir Alexander MacRobert of Douneside Aberdeenshire MRA: MS Section 2 correspondence 9/5/4. 1910. Letter to Sir Alexander of Douneside from Rachael Workman in: *The First Female Fellows and the Status of Women in the Geological Society of London Cynthia* vs. Burek Geological Society Special Publications, Burek, 2009, pp. 373–407.

[iii] The MacRobert Trust. http://www.themacroberttrust.org.uk

Janet Vida Watson

Name and Post Nominals

Janet Vida Watson (1923–1985), FRS, PhD, BSc.

First woman professor, Geology department Imperial College.

First woman president, Geological Society.

Specialism

Precambrian Geology of Scotland, Geochemistry

Career Roles

Research Assistant 1952–1973, Senior Lecturer 1973–1974, Personal Chair in Geology, 1974–1983; Professor Emeritus, Senior Research Fellow 1983–1985; Author of textbooks and papers on geology.

Career Awards

Imperial College Watts Medal; Geological Society of London: Lyell Fund (jointly John Sutton) 1954; Bigsby Medal (Jointly John Sutton) 1965; Lyell Medal 1973; Geologists Association Foulerton Award 1972; Clough Medal Edinburgh Geological Society 1980; Fellow President of Section C British Association (BAAS) 1972; Royal Society Fellow 1979.

Professional Affiliations

Fellow President of Section C British Association (BAAS) 1972; Vice President Geologists Association 1973–1977; National Water Council 1973–1976; Member International Board International Geological Correlation Programme (ICGP) 1977–1980; National Correspondent, Lithosphere Programme (ICSU) 1980–1984; President Geological Society 1982–1984; Ex-officio member, chairman, British National Committee for International Geological Correlation Programme (ICPG), 1980–1981; Assessor to NERC. Council member and Vice President, Royal Society 1983–1985.

Lectures

In Her Name

Watson Forum Project, an initiative in Janet Watson's name to promote women working in mathematical modelling, simulation and programming in the context of maths, physics, earth sciences and engineering.[i]

Quote

What is past is prologue.[ii]

> Janet Watson with her fundamental knowledge and understanding, creative intellect and totally workmanlike approach to pure and applied branches of her science epitomised all that is best in British geology and particularly the Royal School of Mines Imperial College.[iii]

Born

1923, Hampstead London.

Died

29th March 1985, London.

Family

Father: David Meredith Sears (DMS) Watson, Professor of Zoology, vertebrate palaeontologist, comparative biologist, University College

London; Mother: Katherine Margarite Parker, previously an embryological scientist; Siblings: one sister Katharine Mary.

Marriage

1949 John Sutton (1919–1992) geologist Imperial College (1946–1980) Professor (1958–1983), Pro-Rector (1980–1983). Two daughters deceased at birth.

Education

Schooling

South Hampstead High School.

University

Undergraduate

Reading University 1942–1944; Imperial College 1945–1947.

Degree

BSc Biology, and Geology Reading; BSc Geology, Imperial College.

Postgraduate

Imperial College 1945–1949.

Degree

PhD.

Postgraduate Awards/scholarships

Senior Studentship 1851.

Text Discussing Career and Other Comments

Janet's family background was studious and liberal, non-conformist, but fun — Janet kept show-mice and salamanders as a child, showing her leanings towards biology early on.

As both her parents had been involved in academic life, it was clear that this would be the direction that Janet should take too, although she tried a few career directions before settling on hard rock geology at Imperial College.

Janet went up to Reading University just at the start of WWII, there, she read biology and geology, reflecting her father's interests, graduating BSc First Class in general science, biology and geology, in 1944. These subjects meant that a variety of work choices were available to her. She initially worked in dairy research, then she became a teacher in a Bournemouth girls school.

Postwar, Janet decided to specialise in geology. Her father's understanding of the academic geological world in the UK meant that as she wished to research hard rock geology, she should elect to study at Imperial with H.H. Read.[iv] He was then Head of the Geology Department, and a major influence on the direction of Janet's early work, besides being a close colleague of the Geological Survey of Great Britain. During her career, Janet too worked closely with members of the Geological Survey (GS), thus continuing a Royal School of Mines (RSM) tradition from the 19th century, but taking it to a professional level for women. Women had always been involved, but mostly as family of the RSM geology lecturers and GS surveyors. (At that period the lecturers posts and a Geological Survey post were often held by the same person.) Janet, like the early geologist felt there was no substitute for being in the field and observing first hand. She adhered to this principle herself and imbued it in her students. That way verification of theory was confirmed, or led to further investigation. She was also a believer in cross disciplinary working — her original BSc biology knowledge was applied and extended in her geological researches.

At Imperial, Janet took her BSc in two years and graduated BSc first class in 1947. She thought she was the first woman honours graduate in Geology at RSM.

Continuing her studies, she embarked on a PhD under H.H. Read's direction. Read sent Janet and a fellow student, John Sutton, to northwest Scotland to investigate formations in Precambrian geology that required depth of study. Janet and John completed joint theses and publications on this topic, establishing themselves as experts throughout their lives, on the Lewisian complex, and they also married, that year, 1949. Even on honeymoon they worked, writing later about their findings on the Isle of Sark. Their working collaboration continued all their lives, although John's work became more focussed on internal matters as Head of the Geology Department 1964 and Dean of the Royal School of Mines in the 1965. Janet was also sought after nationally and internationally for her depth of knowledge and incisive grasp of geological and related areas as the discipline developed throughout her career. She was instrumental in consolidating research on the geology of Scotland and the Outer Hebrides, leading her research students in this during the 1960s, basing their work on her joint work with John, yet hugely extending it.

Both Janet and John were appointed as staff in the Geology Department, John as Lecturer in 1948, Janet as H.H. Read's Research Assistant, allowing her to make use of her 1851 award which ran to 1952. Janet and Read developed a great working relationship. They were both communicators, and in Janet's case particularly, after she established her research group, with students. Janet was a student favourite for her clear, perspicacious but warmly enthusiastic discussions and mentoring. She was a very popular speaker on the student geology club circuit. She and Read published two student books together: *An Introduction to Geology* in 1962 Vol. 1, 1975 Vol. 2, and *Beginning Geology* in 1966, to great acclaim.

The Geology Department was seen as a particularly male dominated area, and so when Janet arrived and attended lectures as student, and was clearly not ancillary staff as most other women in the department were, she says her arrival fluttered a few dovecotes. This was compounded by her excellent results. Janet found this funny rather than offensive: "I began to feel more and more at home and became more and more engrossed in the problems of regional geology."[v] Though 40 years later, in 1985, things hadn't changed, people expected a woman answering the phone to be the professor's secretary, not the professor she was. By then, she had been a Professor since 1974 and Emeritus professor since her retirement in 1983, having been appointed to a personal Chair.

By the 1970s, she was invited to De la Beche Club dinners and allowed to attend Geological Society dinners following giving a paper. "The best antidotes are a fairly thick skin, a capacity to enjoy the ridiculous and a reasonable faith in your own abilities."[vi] Sound advice even today!

Therefore Janet had an understanding of the position of most women in science, and she organised a Women in Geology meeting in the 1970s under the auspices of the Geological Society.

She has the distinction of being the first woman President of the Geological Society 1982–1984. In this role, Janet continued her reputation as an innovator, improving accommodation and gaining external funding for the Palaeogeographic Atlas and enabling the Petroleum Exploration Society of Great Britain (PESGB) to use the society's building. She also caused a journal Petroleum Geochemistry to be founded and an archive store to be built.

Always at the forefront of new uses of technology, from the most basic — the new development of the word processor to type her papers on, to the total gamma scintillometer used for readings in the wild weather of the Hebrides, and then she embraced the use of the British Geological Survey's electronic Image Analysis and Geographical Information Systems to investigate geochemical, geophysical and geological data. Her work on uranium deposits via this system in 1978 benefited the international community.

Her international influence was further demonstrated by her setting up a joint visit to the Ministry of Geology and Mineral Resources of China by the Royal Society and the Geological Society in 1985 to explore collaboration in geoscientific work. This resulted in three projects spanning four years.

Janet enjoyed her life and the geology which informed so much of it immensely. "It was always difficult to reconcile the eminent professor Janet Vida Watson with the delightful charismatic Janet who was continually surrounded by people seeking her attention."[vii] She was quietly keen to encourage young women to take up geology, writing in her inimitable style in an issue of *GEOlogIC* in January 1985 — she died in March 1985.

She worked continually with John Sutton and published up to her death; on the day before she died, he read a joint paper of theirs; 'Lineaments in the Continental Lithosphere' at a Discussion Meeting at the Royal Society.[viii]

Endnotes

[i] Department of Earth Science Engineering Imperial College. Retrieved from https://www.imperial.ac.uk/engineering/departments/earth-science/news-events/outreach/watson-forum/

[ii] Professor Janet Watson Professorial Inaugural Lecture Imperial College, 1977.

[iii] Jane Plant GEOlogIC 240 1985 Geology Department Imperial College in-house journal.

[iv] Herbert Harold Read (1889–1970) FRS Hon Fic Professor of Geology Imperial College, member Geological Survey of Great Britain 1914–1931. During WWI joined Royal Fusiliers 1914–1917 Campaigns: Galliopoli, Senoussi; Battle of the Somme severely wounded, invalided out of the army); George Herdman Professor of Geology Liverpool University 1931–1938; Professor of Geology Imperial College, 1939–1955.

[v] GEOlogIC 235, 1985.

[vi] *Op cit.*, iii.

[vii] *Op cit.*, iii.

[viii] *Op cit.*, iii.

Sources: A History of the Geology Department Imperial College 1958–1988. Mary Pugh (Department Administrative Staff).

Engineering

Chemical Engineering

Anita Irene Bailey

Name and Post Nominals

Anita Irene Bailey (1926–) PhD.

Fourth woman professor 1976.

First woman professor, Chemical Engineering Department.

Specialism

Interface science, the science of the mineral mica; Chemical Engineering.

Career Roles

Research roles: Smithson Research Fellowship, Royal Society for work at Cavendish Laboratory Physics and Chemistry Group Cambridge University; Newnham College Fellowship; Member, Governing Body 1956–1958, Member, Council 1961–1964, Associate 1971–, Lecturer in Physics 1957–1963, Acting Director of Studies, Physical Sciences M.1961, L.1962; Research Officer King's College London, Biophysics Laboratory 1964–1969; Institut fur Physik und Chemie Grenzflächen Stuttgart 1970–1976[i]; Kodak Professor of Interface Science Imperial College 1976–1991.

Career Awards

Fellowship Newnham College Cambridge 1954–1957; Special issue of the journal Colloids and Surfaces in her honour 1992.[ii]

Lectures

In Her Name

Anita Bailey endowed the Gillian Vaisey Memorial Lecture at Newnham College[iii] to comprise the Public Understanding of Science and Mathematics.

Born

Third March 1926, Wynberg, South Africa.

Family Background

Father: Cecil John Bailey, a Printer; Mother: Irene Maude Wakeman.

Marriage

None.

Education

Schooling

Jeppe High School Johannesburg.

University

Undergraduate

University of Witwatersrand.

Degree

MSc.

Undergraduate Awards/scholarships

Postgraduate

1950–1954 Cambridge.

Degree

Physics PhD.

Postgraduate Awards/scholarships

Witwatersrand Council of Education Scholarship 1950–1952, 1851 Commission Senior Studentship 1954–1956.

Text Discussing Career and Other Comments

Anita Bailey arrived in Britain four years after the end of WWII, in 1949. The country was still undergoing rationing to save scarce resources including food — to 1954 and clothes — just ended in 1949, and where new furniture was of the utility make, solid and basic. A country very different from her native South Africa. Yet she thrived in the laboratories of Cambridge at the Cavendish 1949–1963, and at Newnham College where she held a Fellowship 1954–1957, and much later at Imperial College. Her distinguished career culminated in the award of a personal Chair endowed by Kodak in Surface Science in 1976, thus becoming the first woman professor in Imperial's Chemical Engineering Department. Anita held the Kodak Chair until her retirement in 1991.

Anita was greatly supported by her male colleagues throughout her career, but particularly two who joined her group at Imperial College in the Chemical Engineering Department, B.J. Briscoe in 1979 and P.F. Luckham in 1983. In his preface to the journal issue in her honour, Briscoe remarks on her help to him and his colleague Luckham with their careers.[iv]

Another colleague, visiting Professor at Imperial in the 1980s, David Tabor, remarks upon meeting Anita at the Cavendish in Philip Bowden's

Physics and Chemistry Group in 1950, as it was so unusual to find women in the position of experimentalists. But he continues that she impressed the lab — which was male dominated — with her work and manner, and also her ability to play the piano, and her enjoyment of calisthenics as a form of exercise. Anita was awarded her Cambridge University PhD in 1954. She had been working with Jeof Courteney-Pratt on the versatile mineral mica which she used as model surface (which has uses in the electricity and building industries, besides others). Mica was a substance she made her own, in terms of the huge advances and developments in its capabilities and in the applications she created, and which continued when she left the Cavendish for London in 1963.[v] She worked in the MRC Biophysics Laboratories at King's College in 1964, before going to the Institut fur Physik und Chemie de Grenzflächen Stuttgart Germany, (now The Fraunhofer Institute for Interfacial Engineering and Biotechnology IGB.) Anita returned to London in 1976, to take up the Kodak Chair of Interface Science in the Chemical engineering Department at Imperial.

She introduced the highly successful Post Experience Courses in Colloid and Interface Science.

Her contributions to surface science throughout her career have been original and highly valued as has her friendship and help to her colleagues wherever she worked. Her courage in dealing more recently to her retirement in 1991 with severely failing eyesight is also commended.[vi]

Endnotes

[i] Fraunhofer-Institut für Grenzflächen-und Bioverfahrenstechnik from 1976. Retrieved from https://www.stuttgart.de/en/item/show/337165/1. Accessed 6 July 2016.

[ii] *Colloids and Surfaces* **63**, p. 7, 1992 (Elsevier).

[iii] Gillian Vaisey excelled at mathematics and was a Research Fellow at Newnham from 1954 to 1957, the same time as Anita held her Fellowship. The lecture was endowed in her memory by Professor Anita Bailey — alumna and former Senior member of the College — with the subject matter to comprise 'The Public Understanding of Mathematics and Science,' 2015.

[iv] B.J. Briscoe *Colloids and Surfaces* **65**(1), pp. 9–17, 1992 (Elsevier. In Honour of Anita Bailey).

[v] D. Tabor *ibid.*, p. 9.

[vi] D. Tabor *ibid.*,

Dame Julia Stretton Higgins

Name and Post Nominals

Dame Julia Stretton Higgins (1942–)
Professor, DBE, CBE, FRS, FREng, FIM, CChem, FRSC, *Chevalier* of the *Légion d'honneur*, DPhil.

Specialism

Polymer Science, Chemical Engineering.

Career Roles

SRC Research Student, Physical Chemistry Laboratory Oxford 1964–1966; Physics teacher, Mexborough Grammar School 1966–1968; SRC Research Fellow, Department of Chemistry University of Manchester 1968–1972; Research Fellow, Centre de Recherche Macromoleculaire CNRS Strasbourg 1972–1973; Physicist, Institut Laue Physicist Institut Laue–Langevin Grenoble — Langevin Grenoble 1973–1976; Successively at Imperial College: Lecturer 1976–1985, Reader 1985–1989, Professor of Polymer Science 1989–2007, Department of Chemical Engineering, Postgraduate Tutor 1989–1994; College Tutor 1990–1993, Dean, City and Guilds College 1993–1997, Director, Graduate School of Engineering and Physical Sciences 2002–2006, Principal, Faculty of Engineering 2003–2007, Emeritus Professor and Senior Research Investigator, Chemical

Engineering Department Faculty of Engineering 2007–, Chairman, Board Trustees, Imperial College London Students Union 2009–2015; President Elect Institute of Physics 2015–2017, President 2017–2019.

Royal Society Associated Organisations

First Chairman of the Steering Committee of the Athena Project, a UK wide government funded activity to improve recruitment, retention and advancement of women in academic posts in Science, Technology and Medicine.[i] Chaired the Advisory Committee on Mathematics Education (ACME) 2008–2012.[ii]

Career Awards

UK Resource Centre's (UKRC) one of six Women of Outstanding Achievement in Science, Engineering and Technology 2010.

Fellowships

Honorary Fellow, Somerville College Oxford; FRSC, Royal Society of Chemistry; FCGI, City and Guilds of London Institute; FIM, Institute of Metallurgy; FInstP, Institute of Physics; FIChemE, Institute of Chemical Engineering; FREng, Royal Academy of Engineering; FRS, The Royal Society.

Awards

CBE 1996; DBE 2001.

Professional Affiliations

Council member, 1998 and 1999; Foreign Secretary and Vice President, 2001–2006 Royal Society; Chairman Neutron Scattering Group, Member, Polymer Physics Committee 1987–1993; President Elect, 2015– Institute of Physics; Member, Council for Science and Technology 1980–1984; Chairman, Scientific Affairs Board; Member, Faraday Council 1984–1987 and 1996–1998; Member, Faraday Education Board member 1994–1997;

Member, Chemistry Research Fund Committee 1991–1996; Member, Chemistry Science Advisory Board 1998– Royal Society of Chemistry; Member 1994–2000, Chairman 2003–2007; Council for Central Laboratory of the Research Councils, 1995–2000 EPSRC; Science and Engineering Committee Advisory Council, 1993–1998; Materials Sector Panel Technology Foresight Programme 1994–1998, British Council; Member, Council for Science and Technology 1998–; Foreign Associate 1999 National Academy Engineering USA; Member, Council for the Central Laboratories of the Research Councils 1995–2000; Member, Research Committee, 1998–2002; Member, Research Assessment Exercise Review, 1999; Member, Chemistry Research Assessment Exercise Panel 1999–2000; Member, Chemistry Panel Research Assessment Exercise 2011 HEFCE; Chair, Council 1998–2000 Science and Technology Working Group on Support for Science Teachers; Chairman, Steering Group Athena Project, Patron Athena Swan 2011.

SERC Neutron Beam Research Committee 1979–1983 and 1988–1991; Institut Laue–Langevi Grenoble (ILL) Science Planning Group for Spallation Neutron Source 1977–1985; Chemistry Sub-Committee 1978–1981; Instrument Sub-Committee 1986–1989, Polymers and Composites Committee 1989–1994; Materials Commission 1991–1994; Facilities Commission, 1993–1994; Journal Editorial Advisory Board of Macromolecules 1984–1987 American Chemical Society; Member, Council 1996–1999 Institute of Metallurgy; President, 2002–2003 Institution of Chemical Engineers; President-elect, Institute of Physics 2016; President, 2003–2004 British Association for the Advancement of Science; Trustee, 1994– Daphne Jackson Memorial Fellowships; Governor, Godolphin and Latymer School Hammersmith London.

Lectures

Given by Her

Founders Prize Lectures, Polymer Physics Group Institute of Physics 2005; Campbell Lecture, University of Southampton Women in Science, Engineering and Technology (WiSET) group 2005; Blackett Lectures Royal Society of Chemistry 2006; Principal Speaker, 2008 World Knowledge Dialogue Symposium; Lady Masson Lecture, University of Melbourne 2008; Selby Lecturer, Australian Academy of Sciences.

Lectures

In Her Name

The Dame Julia Stretton Higgins Lecture, Imperial College.

Quote

> When I asked the question why? in the physics lessons, the teacher was always able to give me an answer.[iii]
> If you are the kind of person who is constantly asking 'why?' and 'what if?' and you want to learn, then perhaps a career in engineering is a logical choice.[iv]
> I was the head of the class [at school] It came as a very great shock at university when I didn't understand everything.[v]
> On working as a post doc: … do you realise they pay me a salary to do what I love doing?[vi]
> Polymer scientists were always orphans in the sense that the polymer science doesn't particularly belong in chemistry or in physics or in engineering, and it was a new subject, so you could find such people anywhere.[vii]

Born

1st July 1942.

Family Background

Her father was George Stretton Downes, a Civil Servant, her mother, Sheilah D. M. Downes née Gavigan, a teacher, two brothers and a sister.

Marriage

1966–1974, Robert Higgins, a musician.

Education

Schooling

Ursuline Convent Prep School, Usurline Convent Grammar School, Wimbledon, London.

University

Undergraduate

Oxford.

Degree

BSc.

Undergraduate Awards/scholarships

Exhibition Physics, Somerville College Oxford.

Postgraduate

Oxford.

Degree

Somerville College, Oxford (MA, DPhil 1968).

Text Discussing Career and other Comments

Born early in World War II, and so into a changing world, where more oppor tunities came into existence for more people, Julia was well placed to reach stellar heights that her career has evidenced. Although her parents were not scientists, Julia and her siblings went into scientific areas; her sister became a maths teacher, one brother an architect and the other went into computing.[iii] Her father was a civil servant at Scotland Yard, who worked his way up to the second highest position, despite not having a university education. Her mother was a teacher in the East End, who gave up work on her marriage — something so many women had to do at that time. Therefore hard work and education were part of the family ethos. Julia read a great deal as a child, but also made things such as most of her clothes, until after her first job!

> *I can't remember anything that would interest anybody about my childhood, in the sense it was a very happy childhood. Nothing dramatic*

happened to us, and we had, as I say, a very good extended family, and I enjoyed that. I mean, I enjoyed being with my cousins and aunts, and so forth.[ix]

She attended the Ursuline Convent Prep School as her parents felt they would have a better chance of getting into the Grammar school on taking the Eleven Plus, Julia is uncertain if this did help, but she and her siblings all went to Grammar school, the Ursuline Grammar School in Wimbledon and her parents were founder members of the parents association, her father became very involved, so was very much part of the children's school life. Julia remembers being very good at maths, so good that she was sent up a class a year early for her age. This was fine from the academic work point of view, but when age regulations about taking the eleven plus prevented her from doing so early, she lost her friends as they all moved on and she had to remain another year.[x] She disliked sports, but enjoyed music and acting. Being Bottom and wearing a donkey's head in "A Midsummer Night's Dream" was a highlight for her.

> She explains ways of working at school: both in maths and in physics, working in a group, so that we helped each other. There's a lot of peer teaching going on, which I've always thought is an excellent way of people learning, because you actually learn as much from trying to explain it to somebody else. If you can do it, you try and explain it to somebody else, you learn a great deal about whatever it is.[xi]
>
> I had a place at Imperial College, and I had a place to do a PhD at Imperial College, which I turned down as well, then eventually I finished up on the staff there[xii] I would work very hard to get to what I wanted. So in that sense hard working, but not hard working because it was a good thing to do, but hard working because I wanted to achieve whatever it was. I was the sort of child that did my homework on Friday night, because I wanted to get it done with.[xiii]

Julia chose to go to Oxford because of the influence of her 'lovely' physics teacher who had been at Oxford, and to Somerville (one of the 5 women's Colleges' she could go to at that time) also because Somerville took science very seriously.[xiv]

At Oxford the physics lectures were not of the best she says, so relied on her weekly tutorial classes and working on her own to build her knowledge. She had an amazing social life, particularly as she discovered Morris and folk dancing one May morning — May 1st being special in Oxford, dawn singing from Magdalen College Tower and traditional dancing and music in the streets.

> In a sense I danced my way through the rest of Oxford, it became a very important part of my life.[xv]

Julia also rowed, joining in the revival of rowing at Somerville, the team competed against a women's eight from Cambridge — not successfully. But never daunted, part of the pleasure of university is the social life, something that takes you into a different realm away from work and gives life another dimensions. Part of the work life balance it is so important to learn as early as possible!

Taking part in music and dancing events, Julia met her now ex-husband who was a musician, in the sense it was an attraction of opposites, which probably was why it didn't last indefinitely.[xvi]

On leaving university, with her PhD thesis still not written up (not to be recommended) Julia and her husband set off to teach in Yorkshire, she physics and he music. They lived in Rotherham and she taught in Mexborough. Neither of them had teacher training, no training in class control or teaching methods, but Julia being a very good teacher acquitted herself well. She demonstrated confidence in her subject knowledge and adequately tackled a new experiment based syllabus, which was even a challenge to experienced teachers. Teaching school children is very different from teaching undergraduates upwards!

Two years of teaching was enough, so what next?

A move to Manchester, and to the University of Manchester, an introduction via her father-in-law, an ICI chemist. It is here that she began her work on neutron scattering of polymers with Geoffrey Allen.[xvii]

She had to travel to Harwell, Oxfordshire a Nuclear Energy Research Establishment[xviii] for some of the work and information which was passed via tapes to the Atlas computer.[xix] There was no internet at the time, the early 1970s, so digital information had to be handled physically.

Similarly Julia says:

'Everything in my thesis had to be hand-drawn. I think I probably got it off a chart recorder eventually…we actually got it back from the original computer on cards and then from the cards you could put it onto a chart recorder.'[xx]

When I became much more senior myself here, my post PhD students and post-docs would write first drafts of things, I would write chunks, particularly the explanatory chunks, but we would both read it through, and then decide if it worked, we'd discuss it. So essentially, and without electronic things, of course all this was being done handwritten and you didn't type things up till quite late on in the process, and you'd be cutting with scissors, and cutting and pasting chunks, you know, if you decided the piece wanted to go elsewhere. It was quite laborious — it's difficult now to remember how laborious it was, working on a paper and moving things around. Now it's so easy with word processors, which ought to mean the papers are better, but I don't think it does always.[xxi]

After four years in Manchester, Julia followed her husband to northern Germany but was soon recruited to a research group in Paris, because of her work on polymers. At the reactor centre at Saclay, working with scientists who knew chemistry but not polymers, but required someone who did have that knowledge. So Julia worked with the Nobel Prize winner Pierre-Gilles de Gennes.[xxii] She had to learn French fast — she had a basic O level understanding and says that even a while afterwards used the French terms in experimental work even when speaking English! She reverted to a semi-student existence, living in a hostel with minimal services at weekends and only a few *emigrés* like herself left, after the more local Europeans went home. Interestingly, Julia notes that there were more women in her field in France than in the UK.

Subsequently she was one of the first British physicists to work at the French/German reactor, adding Britain to the venture as it joined the Common Market (EU today). Julia then worked for the Institut Laue–Langevin, Grenoble. Besides experimentation, her role was to facilitate the experiments of others, to make international contacts and this was not only exciting and stimulating, but broadened her base of scientific colleagues.

Her marriage came to an end, but both her work and social life were very enjoyable in Grenoble, skiing in winter — but due to bruises sustained whilst always in the beginners class, not wearing her short skirts in that season! The international group plus the English expats ate out together and enjoyed each other's company. She had a pleasant apartment and a car, and had begun a slowly developing relationship with her partner, George Stirling whom she had met some years previously whilst working at Harwell.

She was offered a permanent contract after her initial three year stint, but refused:

> as she says: … I did not see myself as somebody who was going to develop the techniques, I wanted to be the person who had the problems.[xxiii]

In 1976, she was invited by a former colleague from her Manchester days, Geoff Allen, to apply for a post in the Chemical Engineering Department at Imperial College where he was then a professor. He required her polymer skills. She decided to take the job, having family and friends still in the UK and not really ever intending to work abroad permanently.[xxiv]

She developed her teaching load gradually and then when Geoff left after a couple of years, she took over the areas he worked on, including research groups and things began to take off for her.

It was at this time, Imperial looked at the way its teachers taught and decided to introduce a teaching diploma at the Institute of Education, and Julia took this up, spending an afternoon a week for two years studying for it. It was a worthwhile experience and subsequently she was involved in developing the first year course, which teaches students how to solve problems. First year students are not used to thinking around topics and finding new solutions — there may be more than one or there may be no correct one, but a way to get things working in an engineering environment.

Julia talks of career 'payback' in these terms:

> in my view if you're going to do research you have to pay for it somehow, so if you want to do it in a lab like Grenoble then you pay for it by

> working on other people's experiments with them when they come in. If you go and do it in industry then you pay for it by doing research in areas you might not be interested in and on topics and being suddenly switched to another area, and if you do it in a university you pay for it by teaching and I thought that the paying for it by teaching was by far the best way of doing it.[xxv]

Julia found there were few women students and one woman professor in her department, Anita Bailey, who was the only woman professor at Imperial at that time. (She was not the first, that was Helen Kemp Porter in 1956: see separate entry.)[xxvi] The numbers did not seem unusual to Julia as she had been used to small numbers of women in science since her undergraduate days — of the eight women who began physics at Somerville with her, four moved to other subjects. At US conferences there were more women of which Julia says:

> it was nice to chat to a few women and we weren't talking about women's things, it was just nice to sit and chat, you know, it was a different feeling. So that was very pleasant to meet those.[xxvii]

From the 1980s onwards Julia's workload increased. Not only were there more students, there was a lot more administrative work and fewer staff at this time in education. (The 1980s were a period of retrenchment ordered by the government.) Julia worked out her teaching and supervision load and her own research, but found it hard to keep up with all concerns, as did everyone else. She was by this time also in demand for external committees, a very important aspect of an academic career.

Julia created her own pressure to be first rate in all she did, publishing papers, working in roles across College and on external committees.

> 'I always wanted to publish the papers because it was a way of completing the piece of work, pulling it together, getting the answer there, getting it out there, so publishing was an important part of the whole process. The pressure to publish would have increased but I don't think I was ever at the level of publishing that I would have particularly felt pressure, I mean I was publishing faster than that would have been a problem and most people in a place like Imperial would be, it wasn't a huge deal.[xxviii]

On being invited to take on a wide pastoral role at Imperial she said:

> 'I'll be a college tutor but not for women, we just have two college tutors'. And I did that for a couple of years and that immediately put me in touch with a whole lot of College, I'd probably done other College wide things before to do with education, but nothing as big as that. Then I became promoted to professor in about 1989, and quite quickly afterwards I was asked if I would be Dean of the City and Guilds[xxix] and that was complete, you know, I mean that really did give me a big college wide role and a lot of work to do. But also much more contact with what you might call the civil service in the College so the Registry and the HR (Human Resources) Department which had to organise this, were then supporting us doing the staff promotion rounds and things, so it was a different interaction with the College and of course with the other Deans and with other senior people and with the other engineering departments.[xxx]

Another College wide role was as Director of the Graduate School in Engineering and Physical Sciences of Imperial College, a part time post, she held between 2002 and 2006.[xxxi] (See also: http://www3.imperial.ac.uk/graduateschool.)

These roles meant she did less science and teaching:

> I never hugely missed doing the experiments but I missed that lovely time of sitting down with a desk full of data and ploughing your way into it and suddenly seeing results come out, you know, patterns from the data, and I miss that. But, you know, you can't do everything. And I do do it a bit, I mean it's still part of the life, I just don't do as much of it. It's a rather peaceful job.[xxxii]

On being elected to the Fellowship of the Royal Society (FRS) one of the highest acknowledgement of status in the scientific world, she says:

> In a certain sense it meant I felt untouchable, I had — I was a scientist, I was clearly a successful scientist and being untouchable isn't quite the right word but I sort of felt, okay, people do believe you're a good scientist you know, you're really there. It also probably encouraged me getting involved in Women in Science and things like that, sort of the

more social aspects which I'd probably put on hold while I was being a scientist. I don't know whether you want to talk about women in science, is that a good point.[xxxiii]

Julia was one of the four Vice Presidents of the Royal Society, as Foreign Secretary 2001–2006 which she describes as 'a wonderful job, the office is part of running Royal Society strategy'.[xxxiv]

Vice Chairman of Vision for science and mathematics education 5–19 project under the Chairmanship of Sir Martin Taylor.[xxxv] This committee is seeking contributions in shaping the scientific education world to be agile and viable in 2030! A long way ahead, so the more input from people in education the better. Julia cites this marshalling of information as hard work, but vital to make the report as visionary as possible and presented in a way which wouldn't be instantly dismissed.

Currently Julia is Chairman of Project BIS Diversity in Scientific Workforce, due to report in 2015. This work is looking at the roles of women. The committee she chairs steers the project team.

Although Julia is now Emeritus Professor she retains a high profile role at Imperial College both in science and in women in science matters. Wisely however, she does manage to escape to the last vestiges of summer in Australia but the is engaged in even more work through her relationship with the University of Melbourne.

When, in her interview with the British Library she states "I was incredibly lucky with a few basic choices in my career and, you know, I've thoroughly enjoyed most of it", she spoke in the past tense too soon — her career is certainly not at an end yet![xxxvi] In 2015, the Institute of Physics unanimously elected Julia President-elect for a two-year term until 30 September 2017, then as President from 1 October 2017 until 30 September 2019.

Endnotes

[i] See Athena Swan.

[ii] An independent committee which acts as a single voice for the *UK* mathematical community, seeking to improve the quality of education in schools and colleges. Retrieved from http://www.acme-uk.org/home.

iii British Council Science personalities Julia Higgins.

iv *Ibid.*

v British Library (BL) Voices of Science, Julia Higgins Interview Transcript July–December 2011. Retrieved from http://sounds.bl.uk/Oral-history/Science/021M-C1379X0055XX-0009V0.BL Julia Higgins, p. 21, C1379/55 Track 2.

vi BL Julia Higgins, p. 92, C1379/55 Track 4 p. 95.

vii BL Julia Higgins, p. 211, C1379/55 Track 8 p. 214.

viii BL Julia Higgins, p. 16, C1379/55 Track 1.

ix BL Julia Higgins, p. 12, Julia Higgins p. 9 C1379/55 Track 1.

x BL Julia Higgins, p. 19, C1379/55 Track 2.

xi BL Julia Higgins, p. 32, C1379/55 Track 2.

xii BL Julia Higgins, p. 38, C1379/55 Track 2.

xiii BL Julia Higgins, p. 39, C1379/55 Track 2.

xiv BL Julia Higgins, p. 44, C1379/55.

xv BL Julia Higgins, p. 49, C1379/55 Track 3.

xvi BL Julia Higgins, p. 53, C1379/55 Track 3.

xvii BL Julia Higgins, p. 92, C1379/55 Track 4G Sir Geoffrey Allen Chemist b. 1928 FRS, FREng.

xviii BL Julia Higgins, p. 94, C1379/55 Track 4.

xix *Ibid.*, Atlas 1st Supercomputer collaborative design between Manchester University and the electronic/electrical companies Ferranti and Plessey. Retrieved from https://en.wikipedia.org/wiki/Atlas_(computer).

xx BL Julia Higgins, p. 75, C1379/55 Track 4.

xxi BL Julia Higgins, pp. 112–113, C1379/55 Track 5.

xxii Pierre-Gilles de Gennes 1932–2007 French Physicist, Nobel Prize Winner for Physics 1991.

xxiii BL Julia Higgins, p. 125, C1379/55 Track 5.

xxiv BLJulia Higgins, p. 184, C1379/55 Track 8.

xxv BL Julia Higgins, p. 191, C1379/55 Track 8.

xxvi See entry of Helen Kemp Porter.

xxvii BL Julia Higgins, p. 201, C1379/55 Track 8.

xxviii BL Julia Higgins, p. 205, C1379/55 Track 8.

xxix City and Guilds College, a Constituent College of Imperial College.

xxx BL Julia Higgins, p. 208, C1379/55 Track 8.

xxxi Retrieved from http://www3.imperial.ac.uk/graduateschool. The Graduate School supports and enhances the postgraduate student experience. Accessed 20160712.

xxxii BL Julia Higgins, p. 210, p. 213C1379/55 Track 8.

[xxxiii] BL Julia Higgins, p. 219, C1379/55 Track 9 p. 222 Vice President, Royal Society p. 222.

[xxxiv] BL Julia Higgins, p. 221, C1379/55 Track 9.

[xxxv] Retrieved from https://royalsociety.org/topics-policy/projects/vision/ Accessed 20160712.

[xxxvi] BL Julia Higgins, p. 246, C1379/55 Track 9 p. 248.

Kathleen Marjorie De Reuck

Name and Post Nominals

Kathleen Marjorie De Reuck (1929–) DIC, FCGI, InstP, CPhys, MSc, BSc.

Specialism

Chemical Engineering, thermodynamics IUPAC.

Career Roles

Deputy Director 1983–1994; Professorial Research Fellow 1994– Co-author of 12 volumes in the series *International Thermodynamic Tables of the Fluid State.* International Union of Pure and Applied Chemistry (IUPAC), Thermodynamic Tables Project Centre, Imperial College London.

Professional Affiliations

IUPAC Subcommittee on Thermodynamic Data; Organising Committee for International Workshops on Equations of State; ESDU Committee on Physical Data and Reaction Kinetics; Member, Women in Physics Group Institute of Physics; Advisor to the Daphne Jackson Fellowship Trust; Board Member, British Pugwash Group; Trustee, Friends of Charing

Cross Hospital; Member, International Energy Agency, Annex 18 on the Thermophysical Properties of the Environmentally Acceptable Refrigerants; EUROMET Project 202 on Gas Densities; Sixth International IUPAC Workshop on Vapour–Liquid Equilibria in 1-Alkanol Mixtures, Liblice, Czechoslovakia; Physics Section Committee of the British Association for the Advancement of Science.

Born

1929.

Family Background

Kathleen Marjorie De Reuck née Gratwick, Father: Percival John Gratwick; Mother: Kathleen Mary; Siblings: four brothers, John Gratwick OBE (1918–1997) attended Imperial College, career engineering and clothing industry (Empire Stores); Stephen (1924–2014) QC specialism IP, BSc Physics.

Marriage

Anthony V. S. De Reuck (1923–2016) BSc Physics 1951, MSc Chemical Engineering and Applied Chemistry 1956, Chemical Engineering 1957 Imperial College; Economics and Political Science PG University College London; Research Fellow, Head of International Relations Studies, Surrey University. Area of work: Peace and Conflict Research; Board Member British Pugwash Group.

Education

University

Undergraduate

Imperial College, Physics (Special).

Degree

BSc 1951.

Undergraduate Awards/scholarships

Physics and Mathematics ARCS 1951.

Postgraduate

Imperial College.

Degree

MSc Physics 1956, DIC Chemical Engineering 1956.

Text Discussing Career and Other Comments

Marjorie grew up with four brothers, so was quite prepared to work with male students, but was surprised on attending Imperial on her first day that there were so few women. She said to her mother who had accompanied her that there must have been some mistake, "this is a man's College".

It was no mistake and it wasn't a man's college. Marjorie held her own and got into the swing of the work and the social life. In fact she met her husband, Tony De Reuck very early on in her time at Imperial — but she enjoyed the very innocent attentions of other suitors before succumbing!

Her tutor for her MSc and DIC was Professor D.M. Newitt.

Marjorie joined social groups available to women, and was Secretary of ICWA, elected President for the session 1951–1952. She was also involved in Imperial College Women's Sports Club (ICWSC) This covered a variety of sports, but was hampered then in creating teams, by the small number of women at College. In June 1952, Marjorie was elected Vice President & Junior Treasure of ICWSC.

Her career at Imperial was unusual. She worked for some of the time unpaid! She wrote the history of the Chemical Engineering Department between 1958 and 1960, as part of the series of departmental histories the then Rector, Sir Patrick Linstead, required. She also cowrote that of the Mechanical Engineering Department with Dr E Glaister, 1960 to 1963.

Marjorie joined the IUPAC Thermodynamic Tables Project Centre Group in Chemical Engineering, working with Dr Selby Angus in 1966. The centre had been started only in 1965, so Marjorie was at the forefront of its development. She was responsible for supplying data on the thermophysical properties of industrially important fluids, used in industries such as natural gas power generation, chemical and refrigeration industries. For some of this work, IUPAC was contracted to the Department of Trade and Industry. The IUPAC work has great value to many aspects of science, including the environment. The work was on an international scale, and Marjorie travelled widely to meet colleagues. The Project later became the IUPAC Transport Properties Data Centre, encompassing the IUPAC Thermodynamic Tables Project Centre.

In 1998 Marjorie submitted extensive evidence to the Parliamentary Select Committee on Science and Technology. She argued that there needed to be government investment in Thermophysical Properties Data in the UK.

Having experienced life as a young woman undergraduate seemingly in a man's world, Marjorie was keen to redress the balance as the 1960s went on. She became part of the group of women determined that family life should be a feature of the development of Imperial College and so worked with like-minded people such as Lady Anne Thorne, Sandra Dawson and Dot Griffiths to set up the College Tutor system and the Day Nursery. Besides being early proponents of the work life balance, they did not see being a woman at Imperial should be a problem, but recognised that others perhaps saw women scientists and engineers as a special case. Marjorie and her colleagues wanted to instil a feeling of equality into College, without making too much of the gender issues so much in the press and in feminist movements. Life at Imperial was not to be about that, but about getting on with the job, as equals, in the lab and the research field.

Being involved in Imperial's women's groups meant that she saw and heard first-hand how women felt and how they saw their roles. For academic women Marjorie was involved in AWISE, the Daphne Jackson Trust, Portia, the women scientists' advocacy body founded in 1997 by a group of women at Imperial, of which Marjorie was one, and she remains a Trustee of this dynamic organisation.[i] The Athena movement and the Elsie Widdowson Trust were also interests. She was involved in Imperial College Women's Club which was (and is still) for all women employees and those retired, and the spouses of visiting academics.

Marjorie worked her way up to senior academic level, retiring as a Professorial Research Fellow and for a while worked part time in the IUPAC Centre. She played a large part in the positive development of attitudes to women in science, particularly at Imperial and retains an interest in the way women work in science.

Endnote

[i] Portia's strategy is to help women and men have the same opportunities for engagement and advancement in science, across all science disciplines by: (1) promoting cultures that are collaborative and sensitive to gender issues; (2) ensuring that quality of research and innovation is enhanced by addressing gender issues, where relevant; and (3) showing how new ideas and markets for science knowledge can be created by including women as coowners and cosolvers of problems facing society. This flagship projects include gensSET: gender in science and the gender Summits: Quality Research and Innovation through Equality. Retrieved from: http://portiaweb.org.uk/index.php/about-us. Accessed 13 July 2016.

DEPARTMENT OF CHEMICAL ENGINEERING AND CHEMICAL TECHNOLOGY

STAFF 1984

Back Row A.Maconachie. D.Dugwell. J.Higgins. A.Gunton. P.Amrit. T.Meredith. R.Wallington. D.Adams.C.Long. K.Dougan. P.Cossar. K.Gurney. H.Avis. C.Smith. L.Holt. D.Chadwick. P.Poggioli. J.Farmer. T.Pushman. K.Senkew.

3rd Row M.Brand. P.Tipple. M.Dix. P.Clay. R.Szczepanski. I.Drummond. J.Perkins. M.Tatsis. M.Symons. C.McKeon. T.Stephenson. A.Barnes. R.King. C.Haddow. L.Trew. R.Lewis. T.Agus. R.Kandiyoti.

2nd Row L.Taylor. C.Fletcher. L.Smith. M.deReuck. S.Machletto. J.Whitfield. S.Ortiz. S.Richardson. F.McCourt. G.Maitland. S.Angus. F.Carleton. G.Heeley. R.Wood. A.Dowden.

1st Row H.Hughes. N.Connolly. C.del Cerro. L.Pyle. G.Saville. D.Young. K.Bett. P.Eisenklam. G.Parry. J.Poggioli. A.Jones. H.Michels. B.Briscoe. L.Moulder.

Front Row L.Kershenbaum. A.Cornish. K.Weale. J.Oakley. H.Sawistowski. R. Sargent. A. Bailey. F.Weinberg. M.Streat. W.Wakeham.

Two Professors 5 years apart Julia Higgins Lecturer, Professor 1989, 3rd left back row; Anita Bailey Professor 1976–1991 seated 4th from right. Marjorie De Reuck IUPAC 4th from left 3rd row

Mechanical Engineering

Margaret Fishenden

Name and Post Nominals

Margaret Fishenden (1889–1977), DSc, FInstP, MSc, BSc.

Specialism

Meteorology Air Pollution, Heat Transmission.

Career Roles

Lecturer, University of Manchester 1910–1915; Director, Domestic Heating Research Labs Manchester Corporation 1916–1921; Officer then Scientific Investigator, Department of Scientific and Industrial Research 1922–1932; Honorary Lecturer 1932–1944, Senior Lecturer 1945–1946, Assistant Professor 1947–1955, Imperial College London.

Career Awards

Fellow Institute of Physics.

Awards

Margaret Fishenden Centenary Memorial Prize Imperial College.

Professional Affiliations

Institute of Physics; Chemical Society; Institution of Heating and Ventilation Engineers.

Quote

> … It is always pleasant to recall the lectures of 'Ma' Fishenden. Her information on the theoretical side of heat transfer was always accompanied by [Saunders] experimental data that confirmed the theory.[i]

Born

24th February 1889 Margaret White, Altrincham, Cheshire.

Family Background

Father: Richard William White, Organist; Mother: Sarah Elizabeth White (née Broadbent); Siblings: three.

Marriage

1915 Richard Bertie Fishenden OBE (1954) (1880–1956) Divorced 1932. One son, Richard Martin Fishenden (1917–2014) Physicist worked at AER Harwell; Associate Institute of Physics 1962.[ii]

Education

Schooling

1890–1902, Governess then private school;1902 Manchester Municipal Secondary School; South Kensington Scholarship; Cheshire County Council Scholarship; First Class Matriculation.

University

Undergraduate

University of Manchester 1907–1909.

Degree

BSc Physics, 1st Class Honours.

Undergraduate Awards/scholarships

Higginbottom Scholar.

Postgraduate

1909–1919.

Degrees

MSc 1910; DSc Physics 1919.

Postgraduate Awards/scholarships

1909 Graduate Scholarship; 1910–1911 Elected Beyer Fellow; 1919 Member of Convocation.

Text Discussing Career and Other Comments

Margaret Fishenden was something of a phenomenon, a person who kept herself to herself, but who pushed on the world of physics research in meteorology, heat transmission and air pollution.

She was an early 'green' researcher, looking into ways to reduce pollution by finding methods of more efficient heating for homes and reducing pollution from industry.

Very little information is available on her background. It is known that she was born in Altrincham and had three siblings, that she married at the age of 26 in 1915 and had one son, born in 1917. She divorced in 1932 and was then a single parent, bringing up her young son, not an easy thing to do in that era. It is not possible to speculate on the circumstances of her divorce, but it is remarkable that despite almost suffering a nervous breakdown following it, she did not falter in her work output.

Margaret's early education was at home with governesses and then at a private school. She had an aptitude for academic work which was shown by the award of scholarships throughout her educational career. The first of these was given by Cheshire County Council in 1902 to attend Manchester Municipal Secondary School. The second was a South Kensington Scholarship and the third a Lancashire County Council Scholarship. Her ability was well recognised and well endowed, and she rewarded her funders by obtaining first class matriculation and then getting an undergraduate award to begin her university studies. She became a Higginbottom scholar and attended the University of Manchester from 1907 and obtained a BSc in Physics in two years.

As she began her undergraduate course, the great Physicist and pioneer of nuclear physics, Ernest Rutherford[iii] was just beginning his tenure as Professor of Physics at Manchester University. Therefore, Margaret was studying in very distinguished company — and not only with Rutherford but with Hans William Geiger[iv] and Joseph Edward Petavel.[v] The latter was to support her in her later career. At Manchester University since 190, Petavel held various posts in the Physics Department, as Mechanics lecturer then Meteorology, and Margaret would have taken his classes as a student. As she began to undertake research, they published papers together, (in her then name of Margaret White) such as on Meteorology in 1907.[vi] Petavel became Professor of Engineering and Margaret continued to work in collaboration with him. They both also worked on ventilation in textile mills and weaving sheds with Cecil H Lander,[vii] with whom Margaret was to become a colleague at Imperial College. The work she, Petaval and Lander carried out led to the imposing of regulations on working conditions.

Gaining a Graduate Scholarship, Margaret then worked on her MSc and was awarded this in 1910. She was elected to the Beyer Fellowship of the University in 1910–1911 and worked as a lecturer between then and 1915. In 1916, she went to work for Manchester City Corporation (the City Council). This was quite a time for her: changing her marital status and her job in the course of a year. She had met her husband Richard

Fishenden[viii] whilst a student; he was to become renowned in the printing business. (However they were to divorce in 1932.)

Margaret's lectureship had been in atmospheric pollution and she was in charge of these studies in the Meteorological Department. She continued this speciality in the Air Pollution Advisory Board of the Corporation, heading the research into smoke pollution in Manchester. Between 1916 and 1926 she researched into domestic heating and heat transfer. This was extremely important and influential work, even provoking a response from Marie Stopes,[ix] of welfare and birth control fame. She also investigated atmospheric conditions, continuing work on humidity and ventilation in the textile mills and weaving sheds. Manchester being one of the centres of the textile industry, she lived in the heartland of the industrial revolution, in atmospheric conditions unknown in 21st century Britain — the air she would have known from her birth would have been full of smoke, ash, steam and pollutants from manufacturing and domestic cooking and heating. She was working from personal experience. During this period she gave birth to her son in 1917, and completed her PhD, awarded in 1919. Margaret was truly multitasking, working at her job, family life and at her academic research qualification. Her work for the Corporation was funded by the Department of Scientific and Industrial Research (DSIR), which led her to move to London to work for the DSIR in 1922.

Things then began to happen for Margaret: Petavel proposed her for a Fellowship of the Institute of Physics, to which she was duly elected in 1922.[x] Margaret was at the Fuel Research Division of the DSIR from 1923 where she continued what was to be a career long collaboration with Cecil H. Lander and a new colleague, Owen Alfred Saunders.[xi]

They quickly established themselves as a powerful research team, even moving to Imperial College together, when Lander was appointed to the Chair of Mechanical Engineering in 1932. Between then and 1945, Margaret was Honorary Lecturer in the Department of Mechanical and Motive Power, in charge of the Heat Transmission Research Laboratories. She was promoted to Senior Lecturer in Heat Transmission in 1945, when she also gave lectures to electrical engineering students in Machines, Thermodynamics and Heat Transfer. Another promotion followed in 1947 to Assistant Professor and Reader in Applied Heat in the Mechanical Engineering Department. Including a greater amount of theory in their courses, they altered the methods of mechanical engineering teaching for the better, reducing the divide between science and engineering. The work Margaret and Saunders undertook and their 1932 text book *The Calculation of Heat Transmission for Engineers* was vital to industry and research and was hugely influential. They produced two other books[xii] and several articles. During WWII they worked on contracts for the Ministry of Supply, as did many others at this period. Saunders and Margaret worked so well together that it was apparently difficult to distinguish between them, where ideas originated from. They shared an office, but it was Margaret who gave time to solving the students problems and providing tea when they required a consultation. Hugh Ford (later Sir Hugh) was one of their PhD students who later became an eminent engineer. He held the Chair in Mechanical Engineering at Imperial 1965–1978.[xiii]

Margaret was clearly respected for her work and treated as their equal by the men she worked with, and the students saw no gender difference either. It was really a male dominated world at that time at Imperial and it would have been unusual for a woman to be lecturing to classes of mostly male students. But putting gender aside and simply being very good at her work made Margaret's career.

She was again in an unusual position in this period of time, socially. She was a an academic, divorced and a single mother, to be any of those in the 1930s was enough to stigmatise people, but to be all three, particularly divorced and a single mother, required much support, which she had from her colleagues and great mental strength, although she struggled with this.

Besides her lecturing and research, she was a stalwart of several national committees relating to heat industries, for two of which she was Chairman, again a bold position for a woman at that period.

On her retirement in 1957, she took the surprising step of eschewing anything related to her academic or industrial work. She disappeared as far as those areas were concerned.

Moving to Goring near Reading, Berkshire, she tended her garden and enjoyed wild flowers (something she had in common with another woman engineer, but a Civil Engineer, Letitia Chitty). Margaret continued collecting in the areas of antiques, watercolours and Persian rugs. She was a skilled bridge player, which might be expected given her engineering and mathematical background, so she had quite a lot of social contact.

She died of pneumonia in 1977, but did not have a Christian ceremony as she had no religious beliefs, and donated her body to medicine.

Endnotes

[i] From Imperial College Alumni magazine: "My student apprenticeship with D. Napier and Son from 1946 to 1949 included study of gas turbine technology and heat transfer at Imperial College, and I still have the textbook on heat transfer that was subsequently published by M. Fishenden and O.A. Saunders. It is pleasant to recall the lectures they gave. She was always called "Ma" Fishenden by the students, perhaps because she was slightly older than Saunders. Her information on the theoretical side of heat transfer was always accompanied by his experimental data that confirmed the theory." Alan Macdonald (Mechanical Engineering, 1948) *Imperial Matters* Issue 31 Winter 2007–2008, p. 2.

[ii] Elections to Institute of Physics Membership 2nd October 1962 The Register November 1962 IOP Science. Retrieved from http://iopscience.iop.org/article/10.1088/0031-9112/13/11/022/meta. Accessed 26 June 2016 Richard Martin Fishenden worked at AERE Harwell.

[iii] Ernest Rutherford (1871–1937) 1st Baron Rutherford of Nelson OM, FRS. Retrieved from https://en.wikipedia.org/wiki/Ernest_Rutherford. Accessed 27 June 2016.

[iv] Hans William Geiger (1882–1945)Discovered atomic nucleus, co-inventor the Geiger counter. Retrieved from https://en.wikipedia.org/wiki/Hans_Geiger. Accessed 27 June 2016.

[v] Sir Joseph Edward Petavel KBE; FRS (1873–1936). Retrieved from http://www.mace.manchester.ac.uk/about-us/hall-of-fame/aerospace-engineering/sir-joseph-ernest-petavel/. Accessed 27 June 2016.

[vi] Quarterly Journal Royal Meteorological Society v.34 1908 Miss M. White, T. V. Pbino, and J. E. Petavel, F.B.S. — *Discussion of the Meteorological Observations at the British Kite Stations,* 1906–1907.

[vii] Cecil Howard Lander (1881–1949).

[viii] Richard Bertram Penrose (1880–1956) Graphic Artist/Printer Editor of the Penrose Annual Graphic Arts Review 1934–1956.

[ix] *Nature*, Volume 117, Issue 2949, p. 658 (1926).

[x] The Institute of Physics is a leading scientific society. We are a charitable organisation with a worldwide membership of more than 50,000, working together to advance physics education, research and application. We engage

with policymakers and the general public to develop awareness and understanding of the value of physics and, through IOP Publishing, we are world leaders in professional scientific communications. Retrieved from http://www.iop.org/about/index.html. Accessed 26 June 2016.

[xi] Sir Owen Alfred Saunders (1903–1993), Owen Saunders was educated at Birkbeck College and Trinity College Cambridge, obtaining degrees in both Mathematics and Sciences. He worked at the Fuel Research Station East Greenwich, between 1926 and 1932, where he met two lifelong colleagues, Margaret Fishenden and Cecil Lander who also became staff members at Imperial, Saunders joined Imperial as Lecturer in the Mechanical Engineering Department in 1932. He was appointed the first Clothworkers Reader in Applied Physics in 1937. His interest in aircraft piston engines and then jet engines led to his being seconded to the Ministry of Aircraft Production in 1942 where he was responsible for turbine engine research and worked with Sir Frank Whittle and Sir Harold Roxbee Cox. He returned to Imperial as Professor and Head of the Mechanical Engineering Department in 1946, succeeding Lander. Besides researching engines, Saunders worked with Margaret Fishenden on heat transmission, they jointly produced a book on it in 1932, which was the standard text for many years. Saunders was the Dean of the City and Guilds College and Pro-Rector between 1964 and 1966.On the sudden death of Sir Patrick Linstead in 1966, he was acting Rector until, in 1967, he became Vice Chancellor of the University of London. Retrieved from http://www.imperial.ac.uk/centenary/timeline/1960.shtml. Accessed 26 June 2016.

[xii] An Introduction to Heat Transfer. M. Fishenden and O.A. Saunders Clarendon press 1950.

Simplified heat Transfer Data for Design Calculations G.Brown, M. Fishenden, O.A. Saunders (Part 2 by J.E. Bacon, M. Fishnden and O.A. Saunders) 1948, 1950 British Shipbuilding Research Association Report No 17, No. 54.

Margaret also published alone on fuel and domestic heating from 1920 to 1929.

[xiii] Sir Hugh Ford (1913–2010) Research project: British Iron and Steel Corporation for the study of heat transfer and fluid flow in Cowper Stoves. Retrieved from http://www.imperial.ac.uk/centenary/memories/ford.shtml. Accessed 26 June 2016.

Civil Engineering

Letitia Chitty

Name and Post Nominals

Letitia Chitty (1897–1982), MICE, FRAeS, MA, BA.

First woman to gain first class Honours in Cambridge University Mechanical Engineering Tripos.

First woman awarded the Telford Medal of the Institution of Civil Engineers.

First woman Member, Technical Committee Institution of Civil Engineers.

First woman Fellow of the Aeronautical Society.

Third woman Associate Member, Institution of Civil Engineers.

Specialism

Civil Engineering, stress analysis.

Career Roles

Technical Assistant; Research Assistant; Lecturer Civil Engineering.

Career Awards

Fellow Royal Aeronautical Society; Premium, 1969 Telford Gold Medal Institution of Civil Engineers; 1971 Fellow Imperial College.

Awards in Her name

Letitia Chitty Prize, Civil Engineering Department Imperial College; Letitia Chitty Centenary Memorial Prize, Newnham College Cambridge.

Professional Affiliations

Associate Member 1947; third woman AM Member 1968 ICE.

Lecture

Given by Her

Institution of Civil Engineers Arch Dams Conference 1968.

Quote

> Miss Chitty's work throughout a lifetime of research has been distinguished by its methodical approach, its thoroughness and meticulous accuracy, and a lively appreciation of the needs of the designer.[i]
>
> Letitia was a pioneer — a quiet revolutionary — who has helped advance the cause of women engineers — not by any violent action — but by her example of quiet dedication to scholarship and education.

Born

15th July 1897.

Family

Father: Herbert Chitty Barrister; Mother: Mabel Agatha Bradby; Siblings: four; one brother Christopher; three sisters; Patience Penelope; Jessie Elizabeth; Helen Louisa.

Marriage

None.

Education

Schooling

1903–1916 Private Tuition and Kensington High School.

University

Undergraduate

Newnham College Cambridge University, 1916–1917; 1919–1921 Mathematics Engineering.

Degree

BA, Mechanical Sciences Tripos, first Woman to receive First Class Honours.

Text Discussing Career and Other Comments

Miss Letitia Chitty was a very unusual woman — particularly for her time. Being born in 1897 into a well-to-do family in Kensington, it would not have been expected that she would desire nor have a career in engineering — yet this is what she did! Much of this career was spent at Imperial College. In later life she was interviewed by *Vogue* magazine, and she also published a book on flowers and was warmly known as an eccentric by the end of her life.

Her early life was spent in Kensington, an exclusive part of London, close to Imperial College. Her mother was the daughter of a clergyman, also headmaster of the exclusive Haileybury School, her father was a Barrister at Inner Temple, 1889 and Secretary, 1903. He was also an antiquary and genealogist. It may have been from her father that Letitia inherited her attention to detail and her tenacity in research, but there is nothing in her immediate family background to suggest an engineering career; her brother went into the church and her sister married a clergyman. The family can trace their antecedents to Sir Frederick Pollock Baronet, via Letitia's father, whose father was Joseph William Chitty married to Clara Jessie Pollock.

Halfway through WWI, in 1916, she went up to Cambridge University to study Mathematics. She was admitted to one of the women's Colleges, Newnham. The division between the sexes was very strict in the early 20th century and there were only three women's colleges as opposed to 22 men's Colleges and no mixed colleges. Letitia took Part 1 of the Mathematics Tripos, after which she was allowed to break her studies to undertake war work. This she did at the Admiralty Air Department, where she investigated stress analysis on RAF experimental aircraft. Flying was still a very new thing at that time, and planes had not previously been used seriously in warfare. There was a lot to learn, not only about piloting a plane, but about the technicalities, the mechanics of engines and the way materials reacted to stresses of air, vibration of the propellers, movement, speed and general handling.

Letitia's involvement with Imperial College and subsequent career came about through her work on aircraft. At the Admiralty Air Department Letitia worked with A.J. Sutton Pippard, who became a lifelong colleague and mentor; he became a Professor of Civil engineering at Imperial College and invited her to work for him there. Another mentor was Richard V. Southwell, whom Letitia worked with at the Air Ministry following her graduation. The web of Imperial connections enveloped them all, as Southwell later became a Professor, and later still, Acting Rector, at Imperial. It was not only her connections that brought Letitia to Imperial, but the recognition by others of her astuteness in research, and her capacity to see problems, work out solutions, and to apply her mathematical expertise to the work in a creative, but uncomplicated way. She enjoyed good design and saw how her mathematical skills could help in making a functional structure, such as a bridge, have pleasing aesthetic qualities. She not only worked well with engineers, but also with designers of large structures.

To continue her postwar academic qualifications, Letitia returned to Newnham and completed her studies, but in engineering, not mathematics, so absorbed had she become in this subject whilst on secondment to the Admiralty Air Department. However, she would only have been awarded a certificate giving her the title as women were not admitted to the degree of Bachelor of Science. In 1921, on graduating with First Class Honours in the Mechanical Sciences Tripos, which Letitia has the distinction of being the first woman to do so, she worked at the Air Ministry on the Airship Stressing Panel, following the disaster with Airship R. 38. Airships R.100 and R.101 were then in production, and it was crucial that they were structurally tested and sound. She also researched on the theory of fluid flow and of framed structure. In all this work, she was Research Assistant to R.V. Southwell, until, in 1934, she became a Research Assistant in the Civil Engineering Department at Imperial College.

Here she worked on stresses in a disc wheel under load, solutions for plane stress problems, suspension cables, plastic collapse of cylindrical shells and the propping of an arch. Her major work later was to be on arch dams and on the behaviour of the voussoir type arch, which the Building Research Station particularly requested her for. Her ability was well known within the research establishments. She was again sought after by The Admiralty, investigating stresses in submarine hulls from explosions and by the Ministry of Supply during the Second World War. She took on this work for the duration of the war, besides a new teaching role in 1943, as Lecturer in the Civil Engineering Department, where she was affectionately known as Letty.

That she was sought after by these organisations should have been enough to prove her abilities and contribution to the profession of Civil Engineering. Yet when, in 1946, he wished to propose Letitia for membership of the Institution of Civil Engineers, Pippard had to make a special case.

She had all the right qualifications and experience, but even so the application was a little unusual. Letitia had gone straight into industry and had not undertaken postgraduate work in an academic situation. She had, however proved her academic qualifications over and over again in her practical, or as Pippard emphasised it, her '*experimental*' engineering work. As an '*Alternative to Training*' Pippard noted: "*Experience as Assistant to Corporate members: 1936–1940 Under Prof A.J. Sutton Pippard: In the Office 3 years". In or on works 1 year.*' Pippard says if this had been undertaken at "a Government Research Laboratory it would have been accepted without hesitation by the Institution." She could, under the rules, submit a thesis on her experimental research, but Pippard made "*a special plea*" that instead of her spending time doing this, in view of his collaboration with Letitia, her age (49) and her publications, plus the fact that she had been a recipient of the Institution's Telford premium at least three times by then, that attaching a list of 12 of her published papers would suffice to prove her suitability for election. This application was necessarily supported by three other members of the Institution, one of whom was at Imperial, Cecil L. Fortescue; it was successful!

In 1958, Letitia was one of only five women members of the Institution of Civil Engineers out of a membership of about 16,000, and that year became the first woman member of the Technical Committee of the Institution of Civil Engineers. In 1969, she became the first woman full member of the Institution to receive the Telford Gold Medal for her publications. This was for her publications throughout her career, culminating in her work on dams and her presentation of her research on the entire work of the Arch Dams Committee of the Institution, to their conference in 1968.[ii]

Letitia also became a Fellow of the Royal Aeronautical Society for her work on aircraft structures.

In 1971, she was made a Fellow of Imperial College; at the ceremony on the stage in the Albert Hall where all Imperial's degree and award ceremonies are held, she was described as "visibly shy" by the reporter in the student magazine *Felix.*[iii] She wasn't a shy character as far as her work was concerned, but she didn't like the limelight in person and she preferred a tutorial situation to lecturing. Her eulogist at her memorial service, Professor John Munro recalls her lectures on Elasticity when he was a postgraduate student, and describes her kindliness and shy enthusiasm for her subject. She took endless trouble over research or to assist someone else in theirs. She was completely absorbed in her work, losing track of the hours, never mind the minutes.

A colleague of Letitia's, Michael Carlile[iv] remembers her eccentricities, in later life, such as carrying around a large shopping basket with a cat in it, sometimes it shared this with her knitting. She continued to attend the then annual University Teachers Christmas party for the children of staff on her own, long after her nieces and nephews had outgrown them; Letitia was still hugely entertained by the conjuring acts.

Yet, she would then attend the Friday evening soirées at the Royal Institution and be fully engaged with the scientific discourse given at the meeting. She was fluent in seven languages, travelled greatly and had a wide interest in the arts and history, besides botany and was a great draughtsman. She didn't marry because, first of all, she was of the generation of women who lost their boyfriends/fiancés/husbands to the fighting of the WWI and then she became absorbed in her work. She had a circle of friends, was close to her family and enjoyed music, including the Beatles. She wrote an unusual book, *Abroad An Alphabet of Flowers*, reflecting her interest and knowledge of nature and travel, illustrating the book with her own drawings of the flowers she describes, and her own anecdotes of the places and circumstances in which she found them.

She never saw any animosity from male colleagues, their only interest (and hers) was to get the job done and she was the best person to do it.

She retired in 1962, but essentially her life didn't change, her greatest interest, numerical methods applied to arch dams absorbed her, and she continued to attend the department and research and write, almost until her death in 1982.

Endnotes

[i] Notes on Miss Letitia Chitty 1969 Archives, Institution of Civil Engineers.

[ii] Pippard Papers Archives Imperial College.

[iii] Felix Thursday, 4 November 1971, p. 8.

[iv] Michael Carlile's personal Communication.

Sources: Imperial College Archives B/Chitty; B/Pippard. *A Hundred Years of Civil Engineering in South Kensington* by Joyce Brown Ed. Civil Engineering Department Imperial College London, 1985.

Civil Engineering / Materials

Constance Fligg Elam Tipper

Name and Post Nominals

Constance Fligg Elam Tipper (1894–1995), DSc.

Specialism

Metallurgy; Crystallography, Strength of Materials, Fractures in iron and steel. Developed new science of strength of metals.

First to use scanning electron microscopy (SEM) in metallurgical research, developed the discipline of strength of materials.

First woman full member, University of Cambridge Faculty of Engineering, one of the first women to gain Natural Science Tripos at Cambridge.

Career Roles

Metallurgical Lab at the National Physical Laboratory 1910–1917; Research Assistant to Professor Sir Harold Carpenter 1917–1929 at Imperial College, Royal School of Mines and Research Assistant, concurrently at The Davy–Faraday Laboratory of The Royal Institution London, The Cavendish Laboratory Cambridge and with Professor Geoffrey Taylor at Cambridge University until 1929, moving to Cambridge permanently as a Researcher, Department of Mechanical Engineering undertook some teaching; Visiting Researcher, Mechanical Division, Naval Research Laboratory Washington DC 1940s; Reader and a full member of Cambridge University, Faculty of Engineering 1949.

Career Awards

Frecheville Fellowship 1921–1923; Royal Society Armourers and Braziers Fellowship 1924–1929; Newnham College Fellowship 1931–1932; Beilby Memorial Prize Institute of Metals 1933; Leverhulme Trust Research Fellowship 1936–1938; Associate Fellow Newnham College 1947–1949; Fellow Imperial College 1963. Sweet chestnut tree planted in the grounds of Newnham College on her 100th birthday.

Awards in Her Name

Constance Fligg Tipper Centenary Memorial Prize Imperial College 2007[i]; ICF Constance Tipper Silver Medal 2013.[ii]

Professional Affiliations

Institute of Metals 1917.

Lectures

Given by Her

Bakerian Lecture Royal Society 1923.

Quote

> I never expected any such honour to be conferred upon me as that of becoming a Fellow of Imperial College, or one that could be more appreciated.[iii]

Born

16th February 1894, Constance Elam Fligg, New Barnet, Hertfordshire.

Family Background

Father: William Henry Elam, Trained at Charing Cross Medical School, General Surgeon and Medical Officer Barnet area; Mother: Lydia Coombes; Sibling: one brother, John, an orthopaedic surgeon.

Marriage

1928 George Howlett Tipper (1881–1947) (Born in Kendal, close to Penrith and Constance Tipper's home on her retirement.) Geologist, Director, Indian Geological Survey retired 1927. An expert on mica he undertook arduous war work for the Ministry of Supply and Mica Control (postwar he became Director of this organisation and Consultant to the Board of Trade). For the British Raw Materials Mission to Washington DC, crossing the Atlantic four times (i.e. eight dangerous journeys under war time conditions,) and flying to India, Australia, New Zealand and Brazil, which took a toll on his health, along with bouts of malaria contracted whilst in India.[iv]

Education

Schooling

Felix School Southwold.

University

Undergraduate

Newnham College Cambridge.

Degree

BSc.

Postgraduate

University of London; University of Cambridge.

Degree

DSc 1929; ScD 1949.

Text Discussing Career and Other Comments

Constance Fligg Elam Tipper, made her career mark by establishing the reason for the breaking up of the so called Liberty ships in mid ocean in freezing conditions during WWII. This was not due to the system of welding that had been introduced to speed up the production of these essential supply ships, as would have seemed likely, but was due to a form of metal fatigue: brittle fracture. She established a test for the strength of metals, still in use, called the Tipper Test. Her expertise was well known in the engineering Department at Cambridge, hence Professor John Baker, who was working for the Admiralty Ship Welding Committee, asking Constance to take on the investigations.

It was not until 1960 that she was finally persuaded by her long term Cambridge colleague, Professor Geoffrey Ingram Taylor, to write what is a very readable book on this period and the brittle fracture problem entitled: *The Brittle Fracture Story*[v] for whilst this is a report and full of technical detail, it is essentially a story, and demonstrates how practical science and discovery can be made interesting to non-technical readers without compromising the scientific information.

Constance was the first person to use SEM to investigate fractures in metals and she used only the second machine built, that of Charles Oatley[vi] in the 1950s, to investigate the crystal structure of fractured metals.

Her work on what was to be a new science of strength of metals began with Professor Harold Carpenter at Imperial, where she worked as his Research Assistant from 1916 to 1929. Together they investigated crystals in metals, it was to be a field in which Constance led. She even turned her attention to archaeological applications in 1933, investigating coins and objects.[vii] By this time, she had a full time career in the Engineering Department at Cambridge. She was mainly a researcher, but took on teaching and during WWII, ran the Heat Treatment Laboratory.

Despite working with eminent men who highly valued her work, such as Harold Carpenter at Imperial and Geoffrey Taylor at Cambridge, and being sought out by John Baker to investigate the cause of the Liberty ships breaking up, it took Constance 32 years to be awarded a professional academic title, that of Reader, which was in 1949. At the same time she was made a full Fellow of the Engineering Faculty of Cambridge University, a first for a woman.

She wrote many research papers, both collaboratively and independently, signing them C.F. Tipper, which androgynous method very nearly caused an upset to the Royal Society's Dining Club. Constance was to give a Bakerian Lecture[viii] jointly with her coauthor, Professor Geoffrey Taylor, and attend the Society's Dining Club dinner always given in the speakers honour afterwards, except that women were not allowed to attend Club dinners at that time! It was therefore fortunate that she realised the Society's mistake and pointed out she was a woman, thus saving embarrassment at the prestigious meeting, and so her invitation was withdrawn and her consolation prize was only a box of chocolates.[ix] This showing how professional bodies can be very out-of-step with their professional worlds. But, amends were made in 1973, when a commemorative Bakerian Lecture dinner was held by the Dining Club and Constance did attend.[x]

Constance retired in 1960, but undertook consultancy work, mainly in the north-west of England, the Lake District, particularly for the Vickers shipyard in Barrow in Furness, advising on metallurgy in their submarine construction.[xi] She had retired to nearby Penrith, living with her brother at Langwathby. She was an accomplished organist, pianist and artist, and enjoyed gardening.

It was highly appropriate that her long, productive life was recognised by Imperial College awarding her a Fellowship. After the ceremony she wrote to Lady Linstead, the then Rector's wife, thanking her for her kindness and turning "an ordeal into a very pleasant experience."[xii] Lastly, her

Cambridge College, Newnham recognised her life and work, by the planting of a tree in the grounds to mark her 100th birthday in 1994. She died in 1995 aged 101.

Endnotes

i See Imperial College Centenary prizes.

ii See entry for Professor Dame Julia King. Awarded the International Fatigue Congress (ICF) Inaugural Constance Tipper Silver Medal 2013, Forum for Engineering Structural Integrity FESI Bulletin Vol. 7: 2 Autumn 2013.

iii Letter: CF Tipper to SIr Patrick Linstead Rector Imperial College 26th October 1963 Imperial College Archives.

iv *Nature* v. 159 21st June 1947.

v *The Brittle Fracture Story* by C.F. Tipper 1962, Cambridge University Press.

vi Professor Sir Charles William Oatley (1904–1996) Physicist and Electrical Engineer Cambridge University Developed the Scanning Electron Microscope with his PhD students in Cambridge from 1948.

vii Journal Institute of Metals and Metallurgical Abstracts Vol LIII part 7 July 1933. Retrieved from http://delibra.bg.polsl.pl/Content/9117/P-99_VolLIII_Part7_1933_AS.pdf. Accessed 26 April 2015.

viii Henry Baker left funds in 1775 to the Royal Society for an annual lecture in physical sciences, hence Bakerian Lecture, which in 1923 was given by Geoffrey Ingram Taylor and Constance F. Elam on *The Distortion of an Aluminium Crystal during a Tensile Test.*

ix Brian Cathcart, No dinner, but a nice box of chocs Observations on science and sexism. *New Statesman*, Published 16 February 2004.

x Catalogue of the papers and correspondence of Sir Geoffrey Ingram Taylor OM FRS (1886–1975). D.83 1924, 1927, 1961–1975 Tipper Constance F. (Nėe Elam) Trinity College Cambridge.

xi David Aris. Retrieved from http://www.cachalots.org.uk/wp-content/uploads/2011/01/History-of-the-Liberty-Ship-SS-Jeremiah-OBrien.pdf. Accessed 20150424 Every effort has been made to contact the author. Retrieved from: http://www.cachalots.org.uk.

xii C.F. Tipper letter to Lady Linstead 27th December 1963, Archives, Imperial College.

Chemistry

Martha Annie Whiteley

Name and Post Nominals

Martha Annie Whiteley (1866–1956), OBE, FIC, DSc, BSc, ARCS.

First woman to hold a teaching post at Imperial College 1904.

First woman elected to Council of the Chemical Society (1928–1931).

Specialism

Chemistry.

Career Roles

1891 Science Mistress, Wimbledon High School; 1900–1902 Science Lecturer St Gabriel's Training College Camberwell; 1903 Lecturer Physical Chemistry Honours course Royal Holloway College; 1904 Teaching scholar, 1905 Assistant, 1908 Demonstrator, 1914 Lecturer, 1920–1934 Assistant Professor, (equivalent of Reader) Chemistry Department, Imperial College, Royal College of Science.

Career Awards

1912–1916 British Federation of University Women Fellowship; 1918 Fellow Royal institute of Chemistry; 1945 Honorary Fellowship Imperial College.

Awards

OBE 1920.

Professional Affiliations

Chemical Society.

Lectures

Quote

> Here you have no ordinary woman. I know of no one more likely to inspire women students to great things in science than Dr Whiteley.[i]

Born

Chelsea London.

Family Background

Father: William Sedgewick Whiteley (c.1839–1922) born in Sheffield. A house agent, surveyor, later a furniture salesman; Mother: Hannah Bargh (c.1844–1870s), born in Sheffield. Secondly, William married Mary Bargh Turner Clarke (c.1880–c.1911). Siblings: three, Lilian, Edith, Arthur.

Marriage

None. Adopted daughter, Winifred Gertrude Turner (1901–1989) her half-cousin, who married John Gurlitt Caney (1899–1955) in 1924.

(Elder brother of solicitor and executor of Martha Annie Whiteley's will, Gerald Gurlitt.)[ii]

Education

Schooling

Chelsea High School later renamed Kensington High School.

University

Undergraduate

Royal Holloway College.

Degree

BSc (University of London External Exams) 1890 Oxford Honours Moderations Mathematics (Oxford University External Exams) 1891.

Postgraduate

Chemistry Department Royal College of Science 1898–1902 part time.

Degree

DSc. 1902, ARCS 1903.

Text Discussing Career and Other Comments

Martha Annie Whiteley (1866–1956) known as MAW, was the first woman to hold a position on the teaching staff at the Royal College of Science (RCS). Just as we greatly prize her, she prized the institution that became Imperial College. The distinction which she especially prized was "the honorary Fellowship of the Imperial College, which was conferred upon her in 1945, in recognition of her outstanding services to chemical science in general, and to its advancement at the Imperial College in particular"[iii]

MAW came from an unlikely background for a chemistry academic at that period, there not apparently being any immediate family influence in her choice of subject. She remained a stalwart family member throughout her life, adopting a relative as her daughter, as her will reveals. She lived with her family for some of her adult life, but also in Cambridge and kept an apartment in South Kensington.

Both her parents were born in the north of England, but were living in London in 1866 when MAW was born in Chelsea. The family moved to Hammersmith by the time she was 6. After MAW's mother Hannah died in the 1870s, her father married Mary Bargh Turner Clarke in about 1880. They moved across the river to Castelnau, Barnes in the 1890s. The unmarried sister of Mary, Rosamund Turner, lived with the family. William was widowed again by 1911. He had a variety of jobs, being a house agent, and surveyor, then a furniture business when he was well into his 70s. (Imperial College Women's Association, ICWA, bought furniture from him for their accommodation at MAWS' direction, according to the recollections of a member, see ICWA entry.)

After leaving Kensington High School, MAW attended Royal Holloway College Egham Surrey, graduating BSc in 1890, though this wasn't to be the end of her studies, she kept them up part time whilst working.

She obtained Oxford University Honours Moderations Mathematics as an external student in 1891, just as she began her first job as Science Mistress at Wimbledon High School. MAW taught there until 1900. She then moved to St Gabriel's Training College Camberwell as Science Lecturer, leaving in 1902. She took the post of Lecturer on the Physical Chemistry Honours course at her old college, Royal Holloway College, in 1903. As she had been continuing her academic studies part time at the RCS since 1898 during her working career, she gained her DSc in 1902, and her Associateship of the Royal College of Science (ARCS) in 1903. At this point her RCS tutor, Professor William Tilden,[iv] asked her to join the academic staff of the chemistry department. This she did in 1904. Her first role was as a teaching scholar in the Chemistry Department, becoming an Assistant in 1905, a Demonstrator in 1908, a Lecturer in 1914 and Assistant Professor (equivalent of Reader) in 1920, she retired in 1934. But retired only nominally, as she still advised students and colleagues on their work and kept up to date by working on editions of the *Dictionary of Applied Chemistry*. She did some of this work at Cambridge University, which offered her accommodation after she was bombed out of her home in London during WWII. She became Editor-in-Chief of the fourth edition in 1940. She completed this shortly before her death in 1956, aged 90.

Colleagues found MAW a very able and willing collaborator, hence her originally being asked by Professor Sir Edward Thorpe[v] for her assistance as contributor and coeditor for the second edition of the *Dictionary*. She worked on later editions with his name sake, but no relation, Professor Sir Jocelyn Field Thorpe.[vi] With him, she was coauthor of *A Student's Manual of Organic Chemical Analysis* (1925).

Her research output was good and her teaching and lab skills superb. However, during her career, she apparently refused to accept higher positions at Imperial College, preferring instead to continue her chemical research work. It was the consensus amongst her colleagues that she was capable of fulfilling duties at full professorial level. It was typical of her too that she would not always allow her name to be added to collaborative published papers.

Described as firm but kind in her disciplined approach to work, colleagues and to all she took on, MAW gained the respect of students, teachers and academic colleagues alike. Her influence in College spread once she had setup the Imperial College Women's Association as the then Rector Sir Alfred Keogh, requested in 1912 saying of her:

"… No one who knows her can doubt that she is one of the most remarkable personalities either amongst men or women in the country. She has an extraordinary power of expression which in some way enables her to obtain a very remarkable influence in the College, and a charm of manner particularly desirable in a teacher …. Here you have no ordinary woman. I know of no one more likely to inspire women students to great things in science than Dr Whiteley …"[vii]

MAW's influence and "power of engendering enthusiasm amongst the women students" extended to all College Departments. She developed a large pastoral role in College, particularly towards women, using her influence to attempt to combat that perennial problem of lack of loos for women. In 1910 she requested increased cloakroom facilities for women staff and students. However, she was always ready to assist students of both sexes in their work. She took an interest in their academic work, and in their recreation, instituting the Whiteley Cup for squash.

A measure of esteem in which MAW was held by external organisations, is that of the award of one of the five Fellowships given by the British Federation of University Women, between 1912 and 1916.[viii]

In the Great War (WWI) MAW worked with her colleagues on several projects for the war effort. She was officially recorded as a volunteer for this work. The RCS was involved in the synthesis of compounds for drugs and improved methods of producing them. These subsequently went into commercial production, mainly for treatment of the troops. She was also involved in the production of sugars and allied substances. MAW led the Trench Warfare section on gases and, trench ventilation, directing the work of Hilda Usherwood (later Ingold), Christopher Kelk Ingold (1893–1970) and G.A.R. Kon (1892–1951).

MAW developed the lachrymatory gas known as S.K., and the incendiary mixture D.W. (Dr Whiteley). For these services she was appointed Officer of the OBE in 1920.

In her battle to ensure that women's place in the Chemical Society be accepted she was as tenacious as in her researches. In achieving the right of women to be Fellows of the Chemical Society in 1920, she is credited with furthering sex equality in British Science.[ix] She worked with Ida Smedley Maclean (see entry) on women's admission to the Chemical Society. One of MAW's greatest supporters in this was William Tilden. Interestingly, a colleague in the City and Guilds College Chemistry Department, Professor W.E. Armstrong[x] was vehemently opposed to women joining the society and having professional roles. Although, paradoxically he did tutor women.[xi] In 1918 she was elected a Fellow of the Royal Institute of Chemistry, and she was the first woman to be elected a member of the Council of the Chemical Society (1928–1931).

MAW campaigned for women to take science as a degree, particularly after WWI when there were so few women to replace men from scientific positions who were killed or severely wounded serving in the forces.

Although it may not have seemed like it at the time, the RCS Chemistry department was way ahead in its women student numbers and appointment of women as staff. MAW had persuaded Sir Edward Thorpe that three places should be set aside for women students. During MAW's time in the Chemistry Department, (that is as a staff member (1904–1934), but also her emeritus associations lasting until 1956), a number of women who also have entries in this book, would have come under her tutelage, or worked as colleagues. These include, besides Ida Smedley Mclean: Frances Micklethwait, Hilda Judd, Helen Kemp Porter (later first woman Professor at Imperial), Hilda Usherwood, Frances Mary Hamer and Nellie Ivy Fisher. Other colleagues included the wife of Professor H.B. Baker, Muriel and Margaret Carlton, the second woman to hold a teaching post at Imperial. However, this wasn't enough for MAW and she discussed the matter with the College authorities when asked to advertise scholarships for women for study in industrial chemistry. These were being offered by the *Common Cause*, the magazine of the women's suffrage movement. She blamed parents and headmistresses for their lack of awareness of the great demand for college trained women as Works, that is industrial, chemists. In 1916, she wrote a short paper for a careers journal suggesting where opportunities lay entitled *The Different Branches of Scientific Research*.

On her retirement in 1934 she was presented with a book in which was an illuminated address and signatures of those staff and students who had known her at the RCS, beginning with the most senior, including H.E. Armstrong and Sir Richard Gregory.

In her declining years she was cared for by her companion Mrs L. Kingsbury.

A.A. Eldridge one of her colleagues and biographers speaks of her warmly: those who enjoyed the privilege of her hospitality cherish memories of her charm and vivacity, of her gentle humour and of her deep religious convictions. Her abilities remained despite physical disability.[xii]

Endnotes

[i] Sir Alfred Keogh Rector Imperial College 1913 Archives Imperial College B/ Whiteley.

[ii] Winifred Gertrude (Turner) Caney (1901–1989) was a niece of William Sedgwick Whiteley's wife, Mary Bargh Turner. (1848–1911) as Rosamond Turner (1866–1937) was Mary's half-sister and her mother. Therefore a half-first

cousin to Martha Annie Whiteley. Personal communication.

iii Obituary Proceedings Chemical Society 1957 A.A. Eldridge.

iv William Tilden (1842–1926) Professor of Chemistry RCS 1894–1926.

v Sir Edward Thorpe (1845–1925) Professor of Chemistry RCS.

vi Sir Jocelyn Field Thorpe (1872–1940) FRS Professor of Chemistry RCS.

vii *Op cit.*, i.

viii The British Federation of University Women founded in 1907 to afford a means of communication and of united action in matters affecting the interests of women. Ida Smedley Mclean was a founder. If is now the British Federation of Women Graduates. Available at bfwg.org.uk.

xi *Ibid.*

x William Edward Armstrong (1848–1937) Professor of Chemistry City and Guild College (Central Technical College) 1884–1913. Student RCS Chemistry 1865–1867.

xi *Pounding on the Doors: The fight for acceptance of British Women Chemists*, Marelene F. Rayner-Canham and Geoffrey W. Rayner-Canham. Bulletin of the History of Chemistry V. 28 No. 2, pp. 110–119, 2003.

xii *Ibid.*, ii.

IMPERIAL COLLEGE OF SCIENCE AND TECHNOLOGY.

SOUTH KENSINGTON

LONDON

Nov. 29

Letter concerning 'Common Cause' scholarship

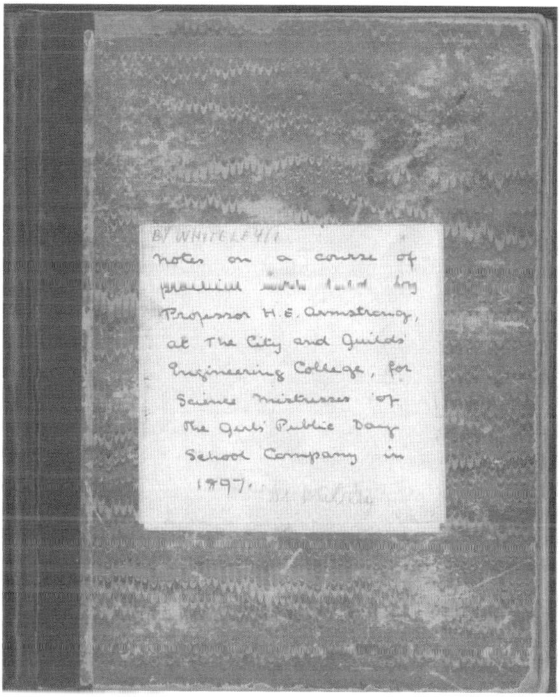

Lecture Notebook Cover 1897

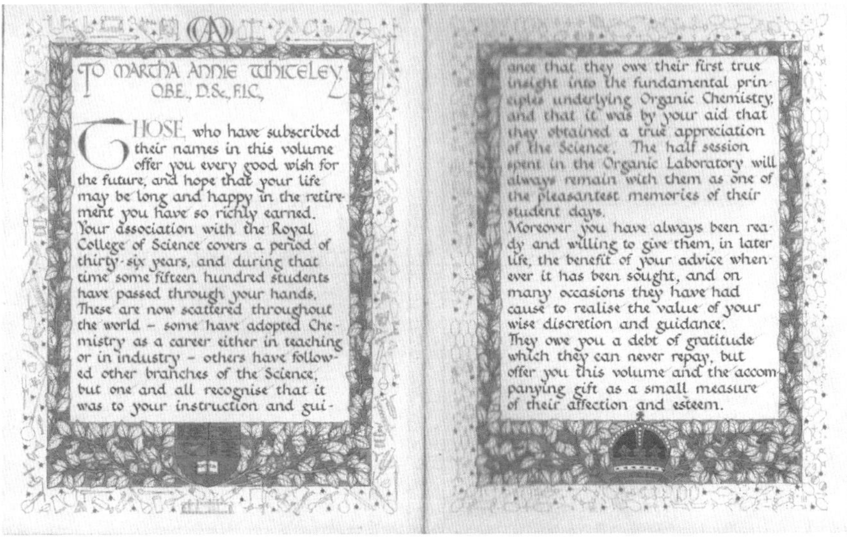

To Martha Annie Whiteley
O.B.E., D.Sc., F.I.C.,

THOSE who have subscribed their names in this volume offer you every good wish for the future, and hope that your life may be long and happy in the retirement you have so richly earned. Your association with the Royal College of Science covers a period of thirty-six years, and during that time some fifteen hundred students have passed through your hands. These are now scattered throughout the world – some have adopted Chemistry as a career either in teaching or in industry – others have followed other branches of the Science; but one and all recognise that it was to your instruction and guidance that they owe their first true insight into the fundamental principles underlying Organic Chemistry, and that it was by your aid that they obtained a true appreciation of the Science. The half session spent in the Organic Laboratory will always remain with them as one of the pleasantest memories of their student days.

Moreover you have always been ready and willing to give them, in later life, the benefit of your advice whenever it has been sought, and on many occasions they have had cause to realise the value of your wise discretion and guidance.

They owe you a debt of gratitude which they can never repay, but offer you this volume and the accompanying gift as a small measure of their affection and esteem.

Martha Annie Whiteley Retirement Book Testimonial

Ida Smedley MacLean

Name and post nominals

Ida Smedley Maclean. From image of Lister Institute Staff, The Wellcome Library.

Ida Smedley MacLean (1877–1944)
DSc, FCS, BA.

Specialism

Biochemistry, fat synthesis and synthesis; acetone production WWI.
First Woman to be awarded a grant from the Chemical Society — £10.
A founder, British Federation of University Women (1907).
One of first three women to hold membership of the Biochemical Society 1913.
Woman Chairman of the Biochemical Society (1927)
Woman to hold membership of the Chemical Society (1920).
Woman to hold a staff position in the Chemistry Department Manchester University: Assistant Lecturer (1906–1910).

Career Roles

Demonstrator Chemistry, Newnham College Cambridge 1903–1904; Research Fellow, Davy–Faraday Laboratory Royal Institution 1904–1906;

Assistant Lecturer, Manchester University 1906–1910; Research Fellow, Research Chemist, Lister Institute of Preventive Medicine 1910–1914 Biochemistry Department 1918–1932; Admiralty Research WWI; DSIR Gas Warfare Department 1914–1918; Cambridge University Appointments Board 1941–1944.

Career Awards

Bathurst Studentship Newnham College 1900–1903; Beit Research Fellowship Lister Institute 1910–1914; Ellen Richards Prize American Association University Women 1915.

Professional Affiliations

Food Investigation Board Committee 1918–1928; Biochemical Society Committee 1925–1928, Chairman 1925–1926; Founder British Federation of University Women 1907–President 1919 and 1929–1935; International Lyceum Clubs for Women Artists and Writers (set up by her sister Constance) subcommittee for universities and science, Chairman International Board 1920s.

Born

14th June 1877, Birmingham.

Died

2nd March 1944, London.

Family

Father: William T. Smedley, Accountant and philanthropist; Mother: Annie Elizabeth Duckworth; Siblings: three. One older sister, Constance, (1876–1941)[i] an author of over 40 books for adults and children, despite severe disability was an artist and theatre company founder; 1909 married the artist Max Armfield (1881–1972).[ii]

Marriage

Hugh MacLean (1879–1957) in 1913. Professor of Medicine at the University of London and St Thomas's Hospital, they met at the Lister Institute. Children: one son born in 1914, and one daughter born in 1917.

Education

Schooling

Taught at home until aged 9; King Edward VI High School for Girls Birmingham.

University

Undergraduate

Newnham College Cambridge; City and Guilds College Imperial College London.

Degree

BSc Natural Sciences.

Undergraduate Awards/scholarships

Gilchrist Scholarship, three years; Bathurst Studentship, Newnham College 1900–1903.

Postgraduate

University City and Guilds College Imperial 1903–1906.

Degree

University of London, DSc 1905.

Text discussing career and other comments

Ida Smedley's schooling was fairly usual for a middle class girl of the period of the late 19th century, that is being taught at home, but less usual as not by a governess, but by her mother. This clearly gave her more time to instil the family attributes of a gifted and studious nature in Ida. Clearly alongside this was instilled a sense of *noblesse oblige*, that is an awareness of the need for the advantaged to assist the disadvantaged in society, and for disadvantage to be put right. But home life was convivial and there were soirees and entertainments besides academic work. It was Ida who developed the scientific bent during her life, although when young she was accomplished in music and acting, and her sister Constance the artistic; both proved their very creative abilities in their chosen areas.

Ida went to King Edward VI High School for Girls at the age of 9, and here her skills in science were encouraged by her tutor Miss Creak, who was one of the first women to graduate from Newnham College Cambridge, and was one of its evangelists. It was therefore natural that Ida should apply to Newnham. She won the Gilchrist Scholarship for gaining four distinctions in the Cambridge University entrance exams.

In 1896, Ida went up to Newnham and took the natural science tripos in 1899, and then began post graduate work at the City and Guilds College with Chemistry Professor H.E. Armstrong and gained a DSc. This is interesting, as Armstrong was not known for his support for women in academic life.

On her appointment as the first woman academic in the Chemistry Department of Manchester University in 1906, Ida would have found a thriving organisation, not only in academic life but in social life. In the chemistry department alone there were many younger academics who all enjoyed cultural activities such as music and theatre. So Manchester with its cultural facilities was an ideal first location for a new academic such as Ida.

In 1910, a new research award was created by the Beit Fellowship Trust. Ida was awarded one of the first of the 10 Beit Memorial Medical Research Fellowships and was the only woman to receive one. She took up a research post at the Lister Institute for Preventive Medicine assisting Arthur Harden[iii] and began research on fats, which was a subject that came to denote her career. Gradually her chemistry became biochemistry.

It was at the Lister that Ida's work as a biochemist began in earnest and this was a relatively new science by name, and one that needed a new society. This was formed in 1911, but, the new committee voted against women members. This was despite Arthur Harden working with Ida, and perhaps something to do with H.E. Armstrong being a founding committee member. Sensibly, the rule was overturned in 1913, and Ida was one of the three women out of seven new members to be elected that year.[iv]

In her profession, Ida was again ahead of her time, although she didn't get a permanent full time post at the Lister until 1932! Her colleagues at the Lister speak of how thorough and competent a researcher she was, how her ideas flowed, and what a pleasure it was to work with her and her enthusiasm for the subject of fats, in carbohydrates and the biochemistry of fats in the body. Ida followed the work of others and with much good science developed it further, thus isolating three omega fatty acids for the first time. She was to become the renowned expert on the subject of fats and their metabolism. Ida was awarded the Ellen Richards Prize of the American Association of University Women related to her research achievements in 1913.

During WWI, 1914–1918, she was engaged in work of national importance by the Admiralty. Much of this work was assisting Chaim (Charles) Weizmann,[v] a Biochemist, in producing acetone through a fermentation process he had devised. Acetone was an important component of artillery. Prewar supplies had come from Germany, during and postwar, Britain had to become self- sufficient in many products, scientific and otherwise.

She produced 35 papers on her subject area, mainly published in the *Biochemical Journal* 1920–1941. She was a just author and much of her work was recognised as collaborative. Ida and her husband, Hugh MacLean, had met at the Lister, and collaborated in their work, even after he left to work at St Thomas's Hospital. They wrote the 1927 edition of *The Lecithins and Allied Substances: Lipins*, together. In 1943, Ida did publish a monograph alone, *The Metabolism of Fat.*

Ida did not confine her ambitions for equality to her professional life, nor to the UK. She advocated women's rights in India too, and for equal welfare payments for men and women, meagre and means tested though they were. Ida was ahead of her time in her ideas of social justice in the 1930s, and also in leading a fulfilling life with a career and a young family.

An early example of a woman continuing with a high powered career and family, Ida managed work, marriage and family happily. Even if she had full time domestic help, which would have been likely in her circumstances, she had to be highly organised to keep everything in peak order. Much as the scientific women of today would do!

Endnotes

[i] Grace Brockington, 'Smedley, (Annie) Constance (1876–1941)', *Oxford Dictionary of National Biography*, Oxford University Press, 2004; online edn, Sept 2013. Retrieved from [http://www.oxforddnb.com.ezproxy2.londonlibrary.co.uk/view/article/76293. Accessed 25 May 2016. Smedley (1876–doi:10.1093/ref:odnb/76293.

[ii] Peyton Skipwith, 'Armfield, Maxwell Ashby (1881–1972)', rev. *Oxford Dictionary of National Biography*, Oxford University Press, 2004. Retrieved from [http://www.oxforddnb.com.ezproxy2.londonlibrary.co.uk/view/article/30765. Accessed 25 May 2016]. Maxwell Ashby Armfield (1881–1972): doi:10.1093/ref:odnb/30765.

[iii] Sir Arthur Harden FRS (1864–1940) Joint Nobel Prize Winner for Chemistry 1929 for investigations into fermentation of sugar and fermentative enzymes, with Hans Karl August Simon von Euler-Chelpin (1873–1964) Stockholm University. Retrieved from https://en.wikipedia.org/wiki/Arthur_Harden. Accessed 25 May 2016.

[iv] History of the Biochemical Society 1911–1986 T.W. Goodwin. Retrieved from http://www.biochemistry.org/AboutUs/Archive/BiochemicalSocietyHistory.aspx. Accessed 25 May 2016.

[v] Retrieved from: https://en.wikipedia.org/wiki/Chaim_Weizmann

Sources: Royal Society of Chemistry 175 Faces of Chemistry celebrating Diversity in Science. Retrieved from http://www.rsc.org/diversity/175-faces/all-faces/dr-ida-smedley. Accessed 25 May 2016.

Mary R. S. Creese, 'Maclean, Ida Smedley (1877–1944)', rev. *Oxford Dictionary of National Biography*, Oxford University Press, 2004; online edn, Oct 2009. Retrieved from [http://www.oxforddnb.com.ezproxy2.londonlibrary.co.uk/view/article/37720. Accessed 25 May 2016. doi:10.1093/ref:odnb/37720.

Nellie Ivy Fisher and Frances Mary Hamer

Name and Post Nominals

Nellie Ivy Fisher (1907–1995)
BSc, PhD.
First woman to head a division of Kodak Australia.

Specialism

Chemical processes for photographic film development sensitising dyes

Joined Imperial

1926.

Education

Imperial College 1926–1929.

Family Background

Father: Francis Frederick Master Jeweller; Mother: Mary Jane née Davis; Siblings: three.[i]

Career Roles

Research Assistant Imperial College; Research Assistant Ilford Research Laboratories; Chemist for Kodak UK; Founder and Director of Kodak Emulsion Labs, Australia.

About

While studying at Imperial, she was President of Imperial College Women's Association from 1928 to 1929. She worked for Ilford Research Laboratories, then Kodak in Harrow and developed expertise in sensitising dyes for photographic films. She undertook further study while working and was awarded a PhD in 1934. She was invited to go out to the Kodak Labs in Australia and worked in Melbourne from 1939. She founded and ran the Kodak Emulsion Laboratory from 1948 to 1962, where she was an inspirational leader. For recreation she was a member of the Melbourne Walking Club.[ii]

At Ilford and Kodak UK She Worked with Another Imperial Alumna, Frances Mary Hamer

They had at least two patents jointly to their names: *Improvements in the manufacture of dyestuffs and intermediates therefor* United Kingdom Patent GB413300 Publication Date1934-07-09 Assignee Kodak Frances Mary Hamer Nellie Ivy Fisher.

Improvements in the manufacture of cyanine and related dyes United Kingdom Patent GB455710 Publication Date 1936-10-21 Assignee Kodak Nellie Ivy Fisher Frances Mary Hamer.

Name and Post Nominals

Frances Mary Hamer (1894–1980) FRPS
First woman elected to Council Royal Photographic Society 1939.[iii]

First woman to be awarded silver progress medal in recognition of her work in photographic chemistry 1962 Royal Photographic Society.[iv]

Specialism

Organic and Photographic Chemist. Spectral sensitising dyes

Joined Imperial

1945.

Education

North London Collegiate School 1903–1913 (Editor science Journal Searchlight 1912)[v] Girton College Cambridge 1913–1916.

Family Background

Father: Chief Medical Officer of Health London; Mother: Agnes née Conan (staff North London Collegiate School with founder Miss Buss, Godmother and namesake to Frances), Sibling: one brother.

Career Roles

As an undergraduate: Chemical Laboratories with Professor William Pope WWI work; Gamble Essay Prize, Scientific Research Fellowship 1921–1924 Cambridge University; Davy Faraday Research Laboratories Royal Institution 1924; Ilford (photographic company); Kodak Ltd Head of Organic Chemistry Research Department, Research Chemist, Consultant; Honorary Lecturer 1945–1959 Imperial College.

Career Awards

Henderson Award 1948, Progress Medal 1962, Honorary Fellowship 1963 Royal Photographic Society

Endnotes

[i] Personal communication.

[ii] Information from Prof Ian Rae University of Melbourne via Marelene and Geoff Rayner-Canham.

[iii] North London Collegiate School *Notable ONLS.* Retrieved from http://www.nlcs.org.uk/767/old-north-londoners/notable-onls. Accessed 9 July 2016.

[iv] *Ibid.*

[v] *Ibid.*

Sources: Imperial College Archives; *Chemistry was their Life Pioneer British Women Chemists 1880–1949*, Marelene Rayner-Canham, Geoff Rayner-Canham.

Edith Hilda Ingold née Usherwood

Name and Post Nominals

Edith Hilda Ingold née Usherwood

Specialism

Organic Chemistry.

Career Roles

Researcher, Demonstrator.

Born

21st May 1898 Catford London.

Died

1988.

Family

Father: Thomas S. Usherwood Mechanical Engineer, Mother: Edith Howarth.

Marriage

Christopher Kelk Ingold, Chemist; three Children: Sylvia 1927, Keith 1929, Dilys 1931. Lived in Mill Hill North London.

Education

Schooling

Girls Grammar School Lewisham; two years private education in Horsham; North London Collegiate School (Awarded a Clothworkers Scholarship).

University

Undergraduate

Royal Holloway College.

Degree

BSc London (Hons Chemistry).

Postgraduate

Imperial College London.

Degree

DIC 1923; PhD; DSc 1925.

Postgraduate Awards/scholarships

DSIR grant.

Text Discussing Career and other Comments

Hilda, as she was known was born into a working class family in Catford south-east London. Her Father Thomas, was fortunate to come under the tutelage of Henry Edward Armstrong at the City and Guilds College

(C&GC), a constituent college of Imperial College. Developing a friendship, Thomas was one of three people in a similar position who gained teaching posts at Armstrong's newly opened Department of Science, Manual[i] and Art at the new site of Christ's Hospital School at Horsham. This was based on Armstrong's theory of heuristic teaching.[ii] Thomas taught engineering in the manual section. This job necessitated a move to Horsham in 1909 and Hilda left the Girls Grammar School Lewisham where she had been enrolled since 1906, and had two years of private study at home. She showed a great aptitude for science and maths, which it was clear should be developed. Possibly due to her father's association with the C&GC, Hilda was awarded a scholarship by the Clothworkers Livery Company to attend North London Collegiate School, which was itself associated with some of the City Companies. This was the girls school founded by the famous Miss Buss in 1850, who felt that girls could and should be educated to the same level as boys. It also operated on similarly heuristic lines to the department her father taught in, so was in fact a natural choice for Hilda. Miss Buss worked hard to develop the stature of the school and her pupils were among the first to sit, albeit unofficially, the Cambridge Local Exams[3] on a par with boys, except that the girls were not named as doing so publicly. The headmistress during Hilda's time was Dr Sophia Bryant, a dynamic and inspirational teacher who was a mathematician, so ideally suited to develop Hilda's scientific talents.

These were further developed when Hilda attended Royal Holloway College to study organic chemistry between 1916 and 1920; her subsidiary subject was physics, which was to lead her research to be innovative in physical organic chemistry and quantum mechanics, an emerging subject at that time. Her heuristic background would have aided her practicality in experimentation.

During her undergraduate course she won 10 academic prizes and also played hockey and tennis, rowed, played the piano and was secretary of the Music Club. All the girls had to undertake a form of social work based at the women's settlement at Southwark and she was for a time the president. She was a Fabian Socialist all her life, again something developed from her home background.[iv]

From Royal Holloway she went to Imperial College to study under the great Martha Annie Whiteley (see separate entry). She received a grant from the DSIR. Originally Edith enrolled for a year, January–December 1921 so as to train in research in organic chemistry, but stayed on and took a PhD and also a DSc. It was then possible to take a PhD in two years, as it was a question of reaching particular research standards and presenting some papers along with a thesis. In Hilda's case, two research papers and a previously published paper along with her thesis which was based on the Tautometric substances. Her particular work around specific heats and unstable substances was unusual and dangerous.[v]

However, that of her future husband Christopher Kelk Ingold was dangerous too — she rescued him when he was overcome by fumes, much to the delighted interest of their fellow students, as it then became clear that they were romantically linked.

And so Hilda married fellow chemistry student Christopher Kelk Ingold in 1923, and with him moved to the University of Leeds where he was appointed Professor of Chemistry. Hilda found work there as a Demonstrator, but unpaid. They continued researching together and produced Annual Reports for the Chemical Society 1924–1928 — these made Christopher's name.

Joint research work continued after the birth of their first child, and continued at University College London (UCL) where Christopher became head of the Chemistry Department 1930. From 1931 with a second child Hilda had to hold back, but assisted with students when Christopher was unwell in 1933. Around 1939 as WWII approached and scientific immigration to the UK was a necessity for some, Hilda and Christopher supported and worked with, if possible some of the refugees, such as Albert Wasserman.[vi] For some years Hilda could not continue chemistry research, but she and Christopher still worked in the department. When UCL was evacuated to Wales on the outbreak of WWII, Hilda took on an administrative role, coordinating the dispersed sections of the department and acting as both departmental, and, Christopher's secretary. This role she continued on the return to London in 1944 until 1946. Both in 1934 and 1944 she and Christopher sent a joint letter to *Nature* relating to their research, so despite the odds against a full research life for Hilda, they did continue to discuss and work together. Hilda showed how adaptable she was by remaining with her research partner Christopher, both as a scientific partnership and as a marriage partnership, even though she was not able, through domestic life, to play as full a part in daily research work as she may have liked to do. But she was known to be inquisitive and bold in her research.

Endnotes

[i] Manual in this context means engineering.

[ii] Heuristic teaching: learning through discovery

[iii] Cambridge Local Exams similar to GCSE system, to assess attainment of school pupils, Girls could not take these officially until 1867. Both Oxford and Cambridge devised non-university exams and these were precursors of current exam boards which set the national exams for school work. Cambridge Assessment. Retrieved from http://www.cambridgeassessment.org.uk/news/how-have-school-exams-changed-over-the-past-150-years/

[iv] W.H. Brock, *The Case of the Poisonous Socks Tales from Chemistry*, Royal Society of Chemistry 2011.

[v] *Ibid.*, p. 225

[vi] *Ibid.*, p. 228

Biology/Botany

Helen Kemp Porter

Name and Post Nominals

Helen Kemp Porter (1899–1897)
FRS, Dsc, DIC, BSc.

First woman professor at Imperial College.

Specialism

Biology; Botany, polysaccharide metabolism in plants, use of chromatography and radioactive tracers.

Career Roles

Research Assistant 1922, Honorary Lecturer Swanley Horticultural College 1930s; Researcher, Reader in Enzymology and Principal Scientific Officer, Research Institute of Plant Physiology 1957; and Head of Unit and First woman Professor, Imperial College 1959.

Career Awards

Fellow Royal Society 1956; Royal Society of Edinburgh; Agricultural Research Council Society Honorary Member 1964.

Awards

Imperial College Huxley Memorial Medal and Prize 1939; Honorary Associateship Royal College of Science 1964; Fellowship Imperial College 1966.

Professional Affiliations

Biochemical Society; Agricultural Research Council; Agricultural Research Council Society; Royal Society; Royal Society of Edinburgh.

Awards in Her Name

Imperial College's new supercomputer was named 'Helen' in her honour in 2015.

Born

10th November 1899 Hale Farnham, Surrey

Died

7th December 1987 Goring, Oxfordshire.

Family Background

Father: George Kemp Archbold, Headmaster; Mother: Caroline Broughton, née Whitehead, a professional singer, Sibling: one sister.

Marriage

1937 William George Porter a physician, died in 1940s, no children; 1962 Arthur St George Joseph McCarthy Huggett (1897–1968); two step daughters.

Education

Schooling

Early: at home; primary and secondary Clifton High School for Girls.

University

Bedford College University of London 1917–1921; Imperial College 1921–1932; Birkbeck College 1922 (evening course); Chelsea Polytechnic 1928.

Undergraduate

Physics and Chemistry.

Degree

BSc (Hons).

Postgraduate

Birkbeck; Imperial College.

Degree

DSc; DIC 1932.

Postgraduate Awards/scholarships

London County Council Free Place renewable for one year 1921.

Text Discussing Career and Other Comments

Helen Archbold was born into a cultured family and had one older sister, their father was a headmaster and their mother had been trained in music in Belgium, which was probably where they met.[i] The family lived in

Farnham, Surrey until moving to Bristol in 1901. When her father became a master at Redcliffe House Clifton. In early childhood Helen had learnt to read, write and also speak French at home, achievements to be expected with her parentage perhaps. From 1906 at the age of 6, Helen attended Clifton High School for Girls leaving aged 17 in 1917. This was a prestigious academic school, but with an enlightened entrance policy for that time, which ignored class origins.

> … The School was also unusual in that it taught an academic curriculum to a demanding level, much to the delight of its pupils who valued "being taught from the heart and the mind" by women who were themselves highly educated. In 1883 the first scholarship to Oxford was gained and in 1905 a chemistry laboratory was built, making the School one of the first to offer chemistry for girls, thereby giving women the opportunity to follow careers in medicine …[ii]

Helen was therefore in the right place to develop her latent talents in chemistry. She was also able to develop her general academic abilities, and explore other science subjects, at which she clearly excelled, and she was encouraged and inspired by one of her teachers.

Summer holidays were spent with her maternal grandparents in Torquay. Although very Victorian in their daily life — prayers at breakfast with the servants present, Helen was able to get out on the beach and explore, sail and fish with her extended family. They were quite exotic, living abroad with many a tale to tell which influenced her later travel and visits to archaeological sites in the middle east. These childhood summer holidays were her "Halcyon days".[iii]

Helen's family life was entirely altered by the outbreak of the WWI. Her father became Joint Head of Aysgarth School[iv] Yorkshire, (1913–1932) investing money in the school, and her mother took up work in London at the National Government Kitchens. This meant income was stretched, holidays were short and cramped in London flats, leading Helen to decide she would always be self- sufficient and ensure she had enough to live on comfortably. Happily, Helen and her sister were able to remain at Clifton High School as boarders due to financial assistance from two of the school governors. Helen was Head of the boarding house 1915–1916 and Captain of the hockey and tennis teams, thus early on demonstrating her leadership skills.

Her academic ability had already been proved, but was firmly established by her gaining distinctions in english and biology in the University of Cambridge Local Exams Syndicate papers.[v]

In 1917, Helen undertook external courses for the London University Matriculation exam, and gained a place at Bedford College for Women in Regent's Park,[vi] and studied physics, mathematics and chemistry, graduating in 1921.

Obtaining a grant from London County Council which covered her costs, she went up to Imperial College in 1922 to undertake postgraduate organic chemistry research.

She studied under the supervision of Assistant Professor Martha Annie Whiteley[vii] who had created a few places for women in the organic chemistry department with agreement of the Head, Professor Joceleyn F. Thorpe.[viii] Dr Whiteley was a rigorous teacher and gave Helen a through grounding in chemical research through work on preparation and study of barbiturate derivatives. Another mentor to Helen was to be Vernon H. Blackman[ix] who supervised her studies in biology, leading to the award of her Diploma of Imperial College (DIC) in 1923 for the thesis: *Nitrogenous Metabolism in the Apple.*

In 1922, Helen then began working with Blackman in the Department of Plant Physiology and Plant Biochemistry, on of the problem of transporting apples — in ships cold storage they turned brown and inedible. The government DSIR instituted a programme to investigate the cold storage of apples which gave Helen the chance to work as part of Blackman's research team, studying the sugars and organic acids in apples. This area of research was new to Helen and not having studied biology since school, she took evening classes in biology at Birkbeck College,[1x] unbeknown to Blackman, but thereby consolidating her knowledge and her position. This is just one example of her ambition and management of her career.

Blackman worked with Cambridge University Department of Biochemistry, necessitating visits to Cambridge by his team and of course, Helen from 1931. In the labs there, talking to distinguished biochemists such as Sir Frederick Gowland Hopkins, Robin Hill and Muriel Onslow,[xi] Helen's professional interest was piqued to such an extent that she enrolled in biochemistry courses at Chelsea Polytechnic College.[xii]

This was an important change in scientific direction for Helen, her outstanding chemical analyses continued, but she added another dimension by encompassing biochemistry and the metabolism of fruit and its growth. She made carbohydrates a particular study in her work on apples, and later in work on grains such as barley.

Again times were changing for Helen due to external circumstances. Affected by the worsening financial situation of the early 1930s, research funding for the project was reduced. However, Helen was an important member of the team and transferred with the Imperial Group to Blackman's Research Institute of Plant Physiology, which included Rothamsted Experimental Laboratories[xiii] and horticultural colleges. This opened up Helen's opportunities. She worked at Rothamsted and also as a Visiting Lecturer at Swanley Horticultural College.[xiv] Her work then led her to the team of Professor F.G. Gregory,[xv] Head of Plant Physiology at Imperial, another important male mentor for her, where she worked on polysaccharide synthesis of starch and fructose.

Another change for Helen in the 1930s was that She married a physician in 1937, William George Porter, who unfortunately died after only a few years. She retained her maiden name and added his surname to that.

Yet another external factor in the 1930s, the outbreak of WWII in 1939, altered her life again; her section moved out of Imperial and she worked at Rothamsted and lodged in Harpenden. But most importantly, she still had a job — which she enjoyed immensely — and funds enabling her to be self — sufficient as she had always determined she would be.

The postwar years were different again, but this time, it was Helen herself who made the change. Boldly, in 1947, she took a year in America, working with the Nobel Prize winning couple Carl and Gerty Coris on carbohydrate metabolism.[xvi]

Returning to Imperial, her work on starch metabolism continued, resulting in 1949 in an invitation to work with Professor Stanley Peat,[xvii] a carbohydrate chemist in Bangor, North Wales, leading to superb results in her research.

The breakthrough she made in her work at Bangor brought her a research grant from the Nuffield Foundation,[xviii] enabling her to set up her own research lab at Imperial in 1949. It was expected that research would continue along the same lines, but Helen was perspicacious enough to see the value of applying new systems to biochemical research, devised in the 1950s, namely: chromatography and radioactive tracers.[xix] Calling on the assistance of colleagues at Mill Hill Research Station[xx] and Amersham,[xxi] she devised equipment to utilise these methods in the study of plant metabolism, using the tobacco plant, enhancing her reputation as a scientist at the forefront of research.

Her election to the Fellowship of the Royal Society in 1956, was further confirmation of her standing as was her promotion to Reader in Enzymology and to Principal Scientific Officer at the Research Institute in Plant Physiology at Imperial in 1957; and in 1959,to being appointed the first woman professor at Imperial College!

Not only did she take on the mantle of Professor, but also the headship of the Unit of Plant Physiology. Helen continued in these roles until her retirement in 1964.

Externally she was concerned with the profession of biochemists and became very active in the Biochemical Society,[xxii] being elected onto its Committee in 1962, and Chairman in 1965. She guided through changes which kept the Society in a good financial position, making it a Limited Company in 1966 and introducing a more relaxed structure, enabling specialist group working.

Helen may have retired from Imperial, but not from active administrative work which she was also superb at by transferring her meticulous and measured research skills to administrative issues. Besides her work with the Biochemical Society, she became Second Secretary to the Agricultural Research Council,[xxiii] swiftly making her way up in this organisation, to Adviser to the Secretary in 1972.

She had married again in 1962, a widowed physiologist, Arthur St George Joseph McCarthy Huggett[xxiv] but she retained the name Porter. Huggett came with a readymade family of two daughters. Helen knew the family, as a girl in the 1920s, Helen, her sister and Ester Killick, who was to become the mother of the two Huggett girls, had taken a canoe trip on the Danube. Helen was an ideal stepmother and even grandmother in due course, despite the death of Huggett in 1968.

In all her dealings Helen was courteous and was a "caring, scholarly and creative person who could get the best out of her students and colleagues and bring her research projects to a satisfactory and productive conclusion. She was always patient but a hard and meticulous worker, who preferred not to suffer fools gladly and was not fond of trivial talk."[xxv]

Later in life, her interests were hill walking, and she developed her lifelong needlework skills to a high degree, creating "tapestries related to scientific photographs that she had seen, some very complex: ... one represented an image that was reflected onto a screen by a point of light in a parabolic mirror. This was made to indicate an example of mathematical catastrophe of the hyperbolic-umbilic type."[xxvi]

As a tribute to Helen, in 2015, Imperial College's new supercomputer was named "Helen"[xxvii]; apt for someone who thought fast and continually broke new ground.

Endnotes

[i] Dictionary of National Biography (2004, Oxford Dictionary of National Biography, ODNB).

[ii] Clifton High School. Retrieved from http://www.cliftonhigh.bristol.sch.uk/Our-History. Accessed 7 July 2016.

[iii] Biographical memoirs of the fellow of the Royal Society Helen Kemp Porter by D.H. Northcote FRS Vol 37, 1991. Published by the Royal Society London.

[iv] Independent Prep School for Boys. Retrieved from http://www.aysgarth-school.com/. Accessed 7 July 2016.

[v] *Op cit.*, iii.

[vi] Bedford College for Women founded in 1849, based in Regent's Park. Became coeducational in the 1960s. Merged with Royal Holloway to become Royal Holloway and Bedford New College (RHBNC) in 1985. Now known as Royal Holloway. See entry for Professor Dorothy Wedderburn who was head of Bedford College and then the merged RHBNC.

[vii] See Martha Annie Whiteley entry.

[viii] Prof Jocelyn Field Thorpe FRS (1872–1940) Head of Chemistry Royal College of Science.

[ix] Vernon Herbert Blackman FRS (1872–1967) Professor of Plant Physiology and Pathology Royal College of Science.

[x] Birkbeck College University of London Founded by George Birkbeck as the London Mechanics Institute in 1823. Admitted women 1830. Became Birkbeck Literary and Scientific Institution" in 1866. At one time known as the only provider of adult part–time university courses. Retrieved from

https://en.wikipedia.org/wiki/Birkbeck,_University_of_London. Accessed 24 May 2016.

[xi] Cambridge University Department of Biochemistry. Retrieved from http://www.bioc.cam.ac.uk/about/history. Accessed 24 May 2016.

Professor Sir Frederick Gowland Hopkins (1861–1947) FRS, OM Nobel Prize for Physiology or Medicine jointly for discovery of Vitamins 1929. Studied chemistry at the Royal School of Mines 1881–1883.

Robin Hill FRS aka Robert Hill Biochemist (1899–1991). Retrieved from https://en.wikipedia.org/wiki/Robin_Hill_%28biochemist%29. Accessed 24 May 2016.

Muriel Onslow (née Wheldale) Biochemist (1880–1932). Retrieved from https://en.wikipedia.org/wiki/Muriel_Wheldale_Onslow. Accessed 24 May 2016.

[xii] Chelsea Polytechnic formed in 1922, from South Western Polytechnic.Throughout the 1930's Chelsea became established as the chief centre in London for classes in biological sciences, pharmacy and geology. Studies in art and metallurgy were also increasingly popular, and the Department of Chemistry began offering courses in dairy science, food and drugs, and refrigeration. Retrieved from http://www.kingscollections.org/exhibitions/archives/studentdays/chelsea-college/chelsea-college-through-the-years/chelsea-polytechnic-1922–1956. Accessed 24 May 2016).

[xiii] Independent research institute founded in 1843 by John Bennet Lawes, the owner of the Rothamsted Estate and, Joseph Henry Gilbert. Retrieved from http://www.rothamsted.ac.uk/about-us/history-rothamsted-research. Accessed 24 May 2016.

[xiv] Swanley Horticultural College 1889–1945 when it merged with Wye College.

[xv] F.G. Gregory FRS (1893–1961) Plant Physiologist Director Plant Research Institute.

[xvi] Carl (1896–1984)and Gerty Cori (1896–1957) Nobel Prize 1947 in Physiology or Medicine with Bernardo Houssay (1887–1971), for their work on glycogen. Gerty Cori was the third woman to be awarded a Nobel Prize in a scientific field. For details see: Retrieved from http://www.faqs.org/health/bios/77/The-Coris.html#ixzz2ie9Jcgy4. Accessed 24 May 2016.

[xvii] Stanley Peat FRS (1902–1969) Professor of Chemistry Bangor University 1948–1969. Retrieved from http://www.bangor.ac.uk/chemistry/research/past_heads.php. Accessed 9 July 2016.

[xviii] The Nuffield Foundation charitable trust established in 1943 by William Morris Lord Nuffield who founded Morris Motors. Funding research and

innovation in education and social policy and Building research capacity in science and social science. Retrieved from http://www.nuffieldfoundation.org/. Accessed 24 May 2016.

[xix] Chromatography: Method of separating substances into their constituent parts; Radioactive tracing: a method of tracing chemical reactions by combining a radioactive isotope with the substance being studied, which reveals the process of reactions.

[xx] Mill Hill National Institute for Medical Research (NIMR), where chromatography was pioneered is now part of the multi institutional Francis Crick Institute at St Pancras London.

[xxi] Amersham is now a company owned by General Electric. Retrieved from https://en.wikipedia.org/wiki/Amersham_plc. Accessed 24 May 2016. Hisorical information on its website states that it had become a national centre for the development and manufacture of radioactive materials for peacetime uses in medicine, scientific research and industry. Renamed The Radiochemical Centre (TRC) in 1949, it became a government owned body managed by the UK Atomic Energy Authority in 1950. Retrieved from http://files.investis.com/amersham/timeline-frame/science.html. Accessed 24 May 2016.

[xxii] The Biochemical Society exists for the advancement of the molecular and cellular biosciences, both as an academic discipline and to promote its impact on areas of science including biotechnology, agriculture, and medicine. Retrieved from http://www.biochemistry.org/AboutUs.aspx. Accessed 24 May 2016.

[xxiii] Agricultural Research Council established by Royal Charter in 1931, continued to 1981 to coordinate research into agriculture with a view to improving agricultural productivity. Retrieved from http://discovery.nationalarchives.gov.uk/details/r/C10600. Accessed 24 May 2016. https://en.wikipedia.org/wiki/Agricultural_and_Food_Research_Council. Accessed 24 May 2016.

[xxiv] Arthur St George Joseph McCarthy Huggett. 1897–1968 *Biographical Memoirs of the Fellows of the Royal Society By F.W.R. Brambell.* November 1, 1970, 16, pp. 342–364. Retrieved from http://rsbm.royalsocietypublishing.org/content/roybiogmem/16/342. Accessed 24 May 2016.

[xxv] *Op cit.*, iii.

[xxvi] *ibid.*, xxv.

[xxvii] Through a competition open to staff, students and alumni, Imperial has chosen a new name for its super computer, following more than one hundred entries. Retrieved from http://www3.imperial.ac.uk/newsandeventspggrp/imperialcollege/newssummary/news_2-10-2015-10-52-38. Accessed 9 July 2016.

Margaret Lacey Biology Department and colleagues. Folk dancing on Lab block Roof Beit Quad, 1920s. Dr Lacey retired in 1958 and enjoyed revisiting Imperial and seeing the changes, as an elderly lady in 1983.

Biochemistry

Elsie May Widdowson

Name and Post Nominals

Elsie May Widdowson (1906–2000)
CH, CBE, FRS.

Specialism

Nutrition; Composition of the body, Composition of food; Food standards.

Career Roles

Researcher King's College Hospital 1933–1938; Department of Medicine Cambridge University 1938–1946; Wuppertal Germany 1946–1949; Department of Medicine Cambridge University 1949–1968; Head MRC Infant Nutrition Research Division Dunn Nutrition Laboratory 1968–1973; Nutrition Society President 1977–1980; Neonatal Society President 1978–1981; British Nutrition Foundation President 1986–1996.

Career Awards

Honorary DSc University of Manchester 1975; FRS 1976; James Spence Medal British Paediatric Association 1981; Rank Prize Funds Prize for Nutrition 1984; Nutricia International Award 1988; Creation of the Elsie Widdowson Laboratory MRC Unit for Human Nutrition Research

Cambridge 1998; The Elsie Widdowson Library in the Food Standards Agency London 2000; First European Nutrition Award 1983; First Edna and Robert Langholz International Nutrition Award American Dietetic Association Foundation 1992.

Awards

Companion of Honour; Commander of the Order of the British Empire.

Awards in Her Name

Imperial College Elsie Widdowson Fellowship Award.

Professional Affiliations

Royal Society, British Nutrition Society; Neonatal Society; Children's Nutrition Research Centre Baylor College of Medicine Houston, Texas; Royal Society of Chemistry.

Lectures

In Her Name

Elsie Widdowson Lectures held by: British Dietectic Association; The Neonatal Society; MRC Human Nutrition Research; King's College London.

Given by Her

Sanderson Wells Lecture University of London 1970; Second Bristol Myers Award for Distinguished Achievement in Nutrition Research 1982; Federation of European Nutrition Societies Boyd Orr Memorial Lecturer 1983; E.V. McCollum International Lecturer in Nutrition and Award 1985; 1986 Atwater Lecturer and Award 1986; Muriel Bell Lecturer Nutrition Society of New Zealand 1989.

Quote

> If your results don't make physiological sense, think and think again! You may have made a mistake (in which case own up to it) or you may

have made a discovery. Above all, treasure your exceptions. You will learn more from them than all the rest of your data.[i]

Tender loving care of children and careful handling of animals may make all the difference to the successful outcome of a carefully planned experiment.[ii]

Born

21st October 1906, Wallington Surry.

Died

14th June 2000, Lake District.

Family Background

Father: Thomas Henry Widdowson Stationer; Mother: Rose née Elphick Seamstress; Sibling: one sister Eva married name Crane, moved to East Dulwich South London.

Marriage

None.

Education

Schooling

Holy Trinity Public Elementary School Wallington: 1912–1918; Wallington County Secondary School September to December 1918; Sydenham County Secondary School 1919–1925.

University

Undergraduate

Imperial College 1925.

Degree

BSc Chemistry and ARCS 1927.

Undergraduate Awards/scholarships

London County Council Scholarship Free Place.

Postgraduate

Imperial College, Courtauld Institute Biochemistry Middlesex Hospital, King's College London.

Degree

DIC Biochemistry 1931; PhD Chemistry Imperial College 1932; DSc 1948. Courtauld Inst BCM.

Text Discussing Career and Other Comments

Elsie May Widdowson was a pioneer of nutrition in the UK. She had a career spanning over 60 years and one professional partner during all this time, Professor Andrew McCance. In a chance encounter at King's College Hospital in 1933, Elsie, using her knowledge from her PhD at Imperial, ensured that McCance corrected an error in his research on carbohydrate values in food stuffs. He then invited her to join his then team of 2 and they spent the rest of their careers researching nutrition together, a high adventure in discovery and self-experimentation!

How could the study of nutrition be called high adventure? Because it was such a new topic and so essential to the well-being of the nation, that it led them to self-experimentation to prove their theories — sometimes leading to alarming consequences — to travel and to meet many of the population who benefited from their work, and also other professionals internationally, including anthropologists[iii] and government ministers who implemented their work nationally, to share knowledge that was vital to survival and development of a fit and healthy population.

Elsie's family proved to be unusual too. Their father was a travelling stationery sundries man, however, both parents encouraged their two daughters to pursue their interests in science. Her sister, Eva (later name Eva Crane), five and a half years younger than her, was a Nuclear Scientist, but became known as an authority on bees!

Elsie attended schools in Wallington until the family moved to East Dulwich and Elsie then cycled to school in Sydenham. Here she excelled in science subjects, but was a good all round scholar, a serious student who had a good influence on others, a pattern which was to continue throughout her life.

She decided to apply to Imperial College rather than a specifically women's college, though she did remark that there were about 100 men and three women in her first year. This did not hold her back. Elsie completed her degree in chemistry in two years, but had to wait a year before it could be awarded. During this time she worked on amino acid research for her MSc. Her professor and mentor was Professor Sammy Schryver.

In 1929 she was appointed as a temporary assistant in biochemistry research. Her first breakthrough came when she started working for the Plant Physiology Department under Helen Archbold (Helen Kemp Porter, see separate entry.) Under a DSIR contract, Helen's work was on the chemistry and physiology of apples and Elsie joined the team. Even before being awarded her DIC, Elsie had published two papers in the *Biochemical Journal*, under the tutelage of Helen, who gave Elsie "my life-long love of research."[iv] The value of both was the increased knowledge they described on the biochemistry of apples. She later stated that she little realised how momentous this was going to be for the whole of her scientific life.[v] In this role she worked on individual carbohydrates throughout the life cycle of the fruit right up to storage. All this was important agriculturally, economically and nutritionally and actually formed the basis of her life's work.

Periodically Elsie had to travel to Swanley in Kent to collect particular varieties of apples, Kent then being 'the garden of England' where there were many orchards; she enjoyed these outings, particularly in the summer. But plants were not her most loved species, her interests lay with human and animal biochemistry.

Therefore when the funding with Helen's team ended, Elsie moved to the Middlesex Hospital Courtauld Institute, to work with Professor E.C. Dodds and undertook research into kidney metabolism. Many years later she wrote that the resulting paper was to her surprise still referred to as "the pioneer work on the subject."[vi] But this work was short lived — research jobs being hard to find, Dodds advised her to take a post graduate diploma in dietetics which was a relatively undeveloped profession, but becoming increasingly important, and this was to prove to be the case for the rest of Elsie's career.

She signed up to what was the first course at King's College of Household and Social Science, University of London. This was where the rest of her career began — in preliminary work she was assigned to the main kitchen to experience cooking on a large scale. Here she met Andrew McCance, her scientific collaborator for the next 60 years. He was experimenting on composition of meat and fish and changes in cooked food, in his own time whilst training as a doctor. Elsie approached him to ask about his work and he introduced her to it — she was able to correct one of his carbohydrate figures in his tables, from her knowledge of working with apples, and they realised there was a huge amount of work they could do together. Simultaneously with her diploma, Elsie worked with Andrew, (who had obtained an MRC grant for her, the MRC was to be another significant factor in her career) on composition of fruits, vegetable and nuts. This work would bring together her work on fruits and Andrew's work on meat and fish, but Elsie realised they would also have to cover cereals, dairy foods, beverages and sweets, in order to make a useful system applicable to dietetics, values for intake of energy and nutrients and general health. This took about 10 years to complete, the first edition of *The Chemical Composition of Foods* being published in 1940.[vii] But it was 10 years well spent, during which Elsie went to America aboard the Aquitania — sea travel being the only option for the general public — another adventure for Elsie. Here she followed up the advice of Margery Abrahams from her dietetic training to meet the people who were responsible for the American Food Tables that British dietetics had been basing its work on, not wholly satisfactorily. Elsie was not satisfied with the way these tables were calculated, she saw that everything needed to be recalculated and exact figures drawn from the investigation to gain an exact understanding of nutritional values in each foodstuff and she and Andrew were vindicated with their British food tables publications.

Food values had previously been calculated as 'man' values for each member of a family, but Elsie realised this needed to change and information on individuals must be collected and set about doing it.[viii]

Elsie was interested in the absorption of foods and their components and gathered data on this via studies in renal problems, salt and minerals such as iron.

Further adventures included self experimentation — injecting herself with a calcium, magnesium and iron which led her to an understanding of mineral metabolism. Later she and Andrew injected each other with strontium, although one batch was contaminated and made them ill for a short period. Such experimentation is not encouraged now, although she did survive her tests, some were not as pain free as others.[ix]

She and Andrew also studied the wartime daily and weekly food ration devised by the government, which was not an exciting prospect as a daily diet, but they proved that it provided all necessary nutrients and energy levels. The only thing they changed was that bread had calcium added to it, because of the increasing incidence of the disease rickets. Elsie and Andrew proved the positive virtues of the diet by sticking to it and then with other later well known scientists (such as Andrew Huxley[x]) taking vigorous exercise in the Lake District. All remained fit, strong and healthy and passed the arduous cycling, walking and climbing situations with no ill effects.

Postwar their research was invaluable in working out dietary requirements for nursing back to health the survivors of the concentration camps and children particularly in the previously occupied countries which had been deprived of food during the war. She made a three year study of nutrition in children's homes in the Wuppertal area of Germany, leading her to conclude that mental health played a part in absorption of nutrients too.

Bread was the staple of any diet, provided all the micronutrients she had scientifically worked out in the recipe, in Elsie's view, one she passed to the authorities at Sandhurst concerned about the effects of their recruits diet.

Animals also came into her dietetic equations as in her study of the milk of elephants and giant pandas in relation to hand rearing of their young, and comparison with cow's milk. She and a PhD student in America, Olav Oftedal, also worked on an important young seal and bear cub nutrition project. He saw Elsie as the only person in the world who would understand his work, and also enable much of it to happen, which she did by obtaining a grant from the Smithsonian Institute, and doing the practical work with him, something she relished.

Elsie was very approachable and had a down to earth attitude, liked things to be uncomplicated and clearly conveyed. She was therefore keen on the straightforward talking by Alistair Cook in his weekly BBC radio talk *Letter from America*. Keen on the medium of radio, she was also a fan of *The Archers*, the long running BBC drama about "country folk". Which incidentally began as a wartime government information programme and continues today, albeit with a much altered emphasis.

People like Elsie never really retire — as long as they have ideas and can research them, they carry on an attitude which we should applaud as age and gender are no arbiters of positive contributions to science. However in 1973, Elsie did take retirement from the MRC, but only moved to Addenbrooke's hospital Cambridge to work in their Department of Investigative Medicine. She remained there until 1988, when she did finally retire from full time work.

For most of her life, Elsie lived in Cambridge, her cottage sheltered her cats, and her garden provided her vegetables, an orchard fruit, and a place for her bees, kept in collaboration with her sister, Eva Crane, who died in Slough in 2007.

The MRC labs in Cambridge were the location of much of her work and in 1998, the new MRC Unit for Human Nutrition Research was named the Elsie Widdowson Laboratory,[xi] she was able to see the initial ground work starting at the start of the building project.

In the year of her death, 2000, a new government agency, the Food Standards Agency was created in London, and the library was named after Elsie, in recognition of her outstanding work on the understanding and development of food standards and the essential part they play in the health of a nation.

Endnotes

[i] Gail Goldberg, *Tales from the Century: Elsie Widdowson and her Eclectic Research*, 30 January 2013. MRC Centenary. Retrieved from http://www.centenary.mrc.ac.uk/news/page/3/. Accessed 16 May 2016.

[ii] Biographical Memoirs, Fellows of the Royal Society 2002, Ashwell, p. 495.

[iii] McCance and Widdowson, *A Scientific Partnership of 60 years. A Commemorative Volume Prepared as a Tribute to the 60 Year Scientific Partnership between Robert Alexander McCance CBE, FRS and Elsie May Widdowson CBE, FRS* ed Margaret Ashwell British Nutrition Foundation, 1993, p. 35.

[iv] *Ibid.*, p. 32.

[v] *Ibid.*, p. 32.

[vi] *Ibid.*, p. 32.

[vii] In its 7th edition 2015 as *McCance and Widdowson Composition of Foods Summary Edition*, Royal Society of Chemistry.

[viii] *Op cit.*, ii, p. 495.

[ix] *Op cit.*, i.

[x] Sir Andrew Huxley Nobel Prize Winner. Retrieved from https://en.wikipedia.org/wiki/Andrew_Huxley. Accessed 29 April 2016.

[xi] MRC Human Nutrition Unit Cambridge. http://www.mrc-hnr.cam.ac.uk/about us/history/dr-elsie-widdowson-ch-cbe-frs/. Accessed 16 May 2015.

Biochemistry/Tissue Engineering

Dame Julia Polak

Name and Post Nominals

Dame Julia Polak (1939–2014)
DBE, MD, DSc, MedSci.

Specialism

Cell Biology, Tissue Engineering, Regenerative Science.

Career Roles

Worked at hospital appointments in Buenos Aires between 1961 and 1967; Assistant Lecturer Royal Postgraduate Medical School London 1968–1970; Professor of Endocrine Pathology 1984; Chair Department of Histochemistry 1992–1997; Director of Tissue Engineering and Regenerative Medicine Centre, Imperial College; The Julia Polak Research Trust.

Career Awards

University of Sheffield Honorary Degree of Doctor of Science during the University's Centenary Year in 2005; Fellow Academy of Medical Sciences 2012; Honorary Doctorate from the University of Computense, Madrid, Spain (1997); Honorary Fellowship of the Association of Clinical

Pathologists (2003); Visiting Professor at The University of Texas Health Science Center at Houston, Texas (2004); Life Time Achievement Award for contributions to Endocrine Pathology 2006.

Medal, Swedish Society of Pathology (1998); The Cable and Wireless Sir Eric-Sharpe Prize for Oncology (1986–1987); Medal of the Society for Endocrinology (1984); Benito de Udaondo Cardiology Prize (for establishing a new way of visualising coronary arteries) (1967); The Ellison–Cliffe Medal, Royal Society of Medicine (2004); Pathology Society Award for long services to Pathology (2004); Feyrter–Oberndorfer Medal and Certificate for ENETS Lifetime Achievement Award Prague (2006); Medal for Valuable Contributions to Medicine by the Royal Society of Medicine; Medal of the Society for Endocrinology; The Professor Dame Julia M. Polak Lecture Imperial College.

Awards

Dame Commander of the Order of the British Empire 2003; Pioneers to the Life of the Nation Buckingham Palace reception with Queen Elizabeth II and Prince Philip 2004.

Professional Affiliations

Society for Endocrinology; Royal College of Pathologists International Academy of Pathology Association of Clinical Pathologists; Scientific Advisory Board of the Imperial College Institute of Biomedical Engineering: Stem Cell Advisory Board Panel of the joint MRC/UKSCF, Science Advisory Board, (October 2005); Panel EPSRC Peer Review College (2006–2009), Panel of the MRC College of Experts (2006–2010); Steering Group of the UK Stem Cell Immunology Programme (2006–); UK National Stem Cells Network Committee (2006–). Council member of the Tissue Engineering Society International and the Academy of Medical Sciences (2002–2005) European Editor of Tissue Engineering until 2004); International Brain Research Organisation (IBRO); New York Academy of Sciences; American Association of Pathologists; Advisor to the Science and Parliament Committees; The Julia Polak Research Trust.

Author of 982 original papers, 118 review articles and Editor/Author of 27 books.

One of the most Highly Cited Researchers in her field.

Lectures

In Her Name

The Professor Dame Julia Polak Lecture Imperial College

Lectures Given by Her

Pathological Society of Great Britain Israel Doniach Lecture[i] 2004; Medal for Erasmus Wilson Distinguished Lecturer at the Royal College of Surgeons; Oliver–Sharpey Lecturer[ii] at the Royal College of Physicians, Ellison–Cliffe Lecturer[iii] at the Royal Society of Medicine.

Quotes

> The *why* of disease is what interested me.[iv]
>
> I seem to have staggered on topics which are at the frontiers: the gut hormone field, nitric oxide then the stem cells or tissue engineering.[v]
>
> Doing research and writing papers about your discoveries is beautiful and you become famous. But in real life you want to save lives and to protect your discoveries, and make money to bring it back to the research.[vi]
>
> All my research is geared to saving human life[vii]
>
> Regenerative medicine — in my unbiased opinion — will rewrite the books of medicine.[viii]
>
> During my career I have seen many positive changes and women are treated almost equally within the profession.[ix]

Born

29th June 1939 Argentina.

Died

11th August 2014 London.

Family Background

Grand Parents had emigrated from Eastern Europe; Father: a Lawyer; Mother: a writer; Siblings: two brothers, one died at age of 32, the other is a lawyer.

Marriage

In 1967 to Daniel Catovsky, Haematologist; three children, one daughter predeceased Julia.

Education

University

Undergraduate

Buenos Aires.

Degree

Physician's certification in 1961.

Postgraduate

London University

Degree

MD 1964; Diploma in Histopathology 1966; DSc 1980.

Text Discussing Career and Other Comments

Julia had a happy childhood family life in Buenos Aires — she enjoyed education and was a natural student, always top of the class, but says 'doing well academically was expected (by her family), education was important to them. It was never an option not to take up a profession' — her original choice of actress was never realised, mostly because she was not good enough, "I was tormenting my family rehearsing my plays in front of them. They were fed up",[x] but neither was it academically rigorous enough to satisfy her families' requirements.

Julia had a famous neuropathologist uncle, Moises Polak,[xi] and although she didn't know him very well, it may be that his influence was relevant in her choice of medicine as a career.

As she studied at university, she realised that she was more academically inclined than clinically inclined, and so took up pathology for the research angle. It was the mechanisms of disease that captured her imagination.

She met her husband Daniel whilst at medical school and they worked alongside each other in research although his specialty is haematology not pathology, there are many links.

Julia and Daniel came to the UK as postgraduate students in 1967, initially for a year, but this was at a period of political, economic and civil unrest in Argentina.[xii] As that situation deteriorated, and their own families were the subject of persecution, their work was going well and they enjoyed the UK, they and their families decided that the best option for them was to remain.

Julia had been awarded a scholarship to work with Professor Everson Pearse,[xiii] a specialist in histochemistry at the Hammersmith Hospital attached to the Royal Post Graduate Medical School, now part of Imperial College School of Medicine. Professor Pearse was working on the chemistry of cells and creating reagents to light up the cells for examination. They were working beyond the boundaries of pathology and chemistry and were making histochemistry a discipline in its own right She says of their work together, "we were unconventional pathologists".[xiv] Julia's remark highlights the huge developments in science since the 1960s — now scientists are encouraged to be unconventional and work in cross — disciplinary ways.

In an interview, Julia explains what pathologists do, suggesting that the public think of pathologists as they appear in television crime roles as the people who do post mortems and come up with times and methods of death. But she says: "pathology is an extremely broad term, including haematologists who look at blood, chemical pathologists who analyse urine and blood and histopathologists who do the diagnosis." These roles involve a great deal of science from various disciplines. "Hard core pathologists do a lot of mechanistic research, as opposed to routine research."[xv]

In her student days she had to prepare samples for demonstration — not a job for the squeamish — and recounts one alarming experience when, putting her hand into a bucket of specimens one day she found it clasped by a disembodied hand that was to be examined! She fled initially, so startled was she, only to gather herself together in her usual objective way and return to her work.[xvi]

Whilst working with Professor Pearse, they together discovered the "gut endocrine system": that the gut produces hormones, and it is not only the glands in the body that produce hormones. This work led her to collaboration with Professor Steve Bloom[xvii] they discovered the gut brain axis — hormones in nerves as neurotransmitters — then a question occurred to them — why would this not be a whole body occurrence? Why should it only be found in the gut? They researched and found a diffuse neuroendocrine system and then asked the question, what will happen in disease? This led to further research into the lung and to Julia meeting a very important figure in her life, Professor Sir Magdi Yacoub,[xviii] the famous heart specialist.

In the meantime, Julia and Professor Bloom had developed a new discipline: immunocytochemistry. She didn't discover the technique but the application, which was new, to the tissues and pathology and ultimately to cells. Although their papers were accepted for publication in rigorous journals such as *Nature*, the profession did not pay much attention to the discipline of immunocytochemistry for some years. She says "it was hard to demonstrate that this was a robust technique and to actually reach the point where the technique *was* robust was hard. But I wasn't afraid of working hard"[xix] Julia is very matter of fact and down to earth. Although imagination and a vision of what is possible must play a part in research, says "you just show very sensibly and with facts. There are no emotions involved in that."[xx]

Julia's enzyme work led to researching nitric oxide, and she discovered that it was produced in the blood vessels, inflammatory cells (microphages) and in neurons; it is a multipurpose enzyme. Her role is not to discover the thing, but to find where it is produced. This then opens the field for research into applications for treatments, such as Viagra.[xxi]

Julia's professional life and her health are very much intertwined. When she was working on pulmonary hypertension, she was diagnosed with the condition herself. It is very rare she says and it had been thought that she might suffer from asthma and all through her life, flying to conferences to give papers and to meetings and having three children, she had been uncomfortable, but carried on. Eventually it became plain that she could no longer carry on as usual. In 1995, she was admitted to hospital and her condition diagnosed. It was then that Sir Magdi played a part in saving her life, by performing a heart lung transplant. *She even presented her own pathology once recovered, and investigated her own lung tissue, "...They were lovely in terms of pathology...but I couldn't imagine how I had managed to breathe...,"*[xxii] thereby proving her complete objectivity in research. People were astounded at the presentation, but her professional curiosity and understanding had to be satisfied, so she did it.[xxiii]

At Sir Magdi's suggestion, her efforts to raise money for Harefield Hospital, were directed to a charity for her own work: The Julia Polak Research Trust, which supports research students.[xxiv]

Was her illness the reason she went into the next phase of her research, tissue engineering, or tissue regeneration? She says maybe but she is always finding new things to be involved in, so It could be, but it is an exciting new field,[xxv] one she perhaps naturally would have tended towards because of her work at the cellular level.

Tissue regeneration is very important to Julia, not only for herself, but because all her research "is geared to saving human life".[xxvi] She says it is a question of mobilising the body's stem cells. How this is to be done, is a matter for research and creating drugs to stimulate the mobilisation but importantly, without altering the cells. There is worldwide interest in this and a range of strategies. But, she says, what is needed is mass production, industrial-scale production of things for clinical application. No one knows how to do it. That's a challenge.[xxvii]

The subject has a way to go as yet and applications are so far limited. Continuing her "maverick" pathology reputation, she started work with an American bioengineer, Professor Larry Hench[xxviii] "he produces the man-made materials, and I'm the one who works on the cells."[xxix] It was Professor Hench who approached her, she describes the encounter: Professor Hench: "— Hello — I'm an engineer. I would like to know whether you can help me to understand the mechanism by which there is bone repair." And I said "Well, we need to use the cells and try to identify what the effect is at the microscopic level in these cells." They have developed research together and are looking into a 3 dimensional approach — tissues talk to each other so it is not just a case of introducing new ones and letting them get on with things — they have to interact peaceably with cells already there.

Julia and Larry set up a spin-out company *Novothera* to protect their intellectual property, a very important thing to do, and to produce funds for the Centre for Tissue Engineering and Regenerative Medicine at Imperial College.[xxx]

Such is the interest in Julia and her life, that a novel has been written about this episode in Julia's life: Intensive Care by Rosemary Friedman who also wrote a play: play *Change of Heart*" inspired by Julia's story.[xxxi]

Julia's family life is of the utmost importance to her — she had three children and although a nanny looked after them by day, she left at 6 pm when Julia and her husband Daniel took over. Getting married and when the children were born were the biggest moments in her life she says.[xxxii]

Her transformational contribution to science is what many will know her for, but family life is important too. Julia and Daniel got the balance right in the distribution of family enhancing tasks and life — not formally worked out, they just happened, for example, she loves cooking, he looks after the car, she deals with computing, he with the phone calls.[xxxiii]

Julia feels women need to be given confidence to succeed in science, and the Imperial College pledge and work on developing women's in careers in academic science will help in some way towards this.[xxxiv] She did not feel she experienced barriers in her career because she was a woman: "I am very thick skinned so I didn't," she says — perhaps women should take a leaf out of her book and gain confidence from her approach?[xxxv]

At first she spent long days in the lab, but as her reputation grew and she had more researchers in her team she says "the more you progress, the more you allow the younger people to do that work. Now I have the status of an elder statesman. I can advise and direct them. They need to develop their own careers.[xxxvi]

Endnotes

[i] The Doniach Lectureship and Award was founded in 2003, in honour of the late Professor Israel Doniach. The objective of the Doniach Award is to celebrate the life-long contribution of a Senior member of the Society to the science of Pathology and to the Scientific activity of the Pathological Society.

[ii] The Oliver–Sharpey lecture was founded by a gift from Dr George Oliver FRCP (1841–1915) of Harrogate, received in 1904, confirmed by deed 1930 — in memory of Professor William Sharpey FRS FRSEd (1802–1880): £2,000 to endow a lecture or prize, in order to promote physiological research by observation and experiment and encourage the application of physiological knowledge to the prevention and cure of disease and the prolongation of life, Royal College of Physicians London. Retrieved from http://old.rcplondon.ac.uk/history-heritage/College-history-new/Historical-lectures/Pages/Oliver-Sharpey-Lecture.aspx.

[iii] The Ellison–Cliffe Lecture is held annually by the Royal Society of Medicine The lectures, which commenced in 1987, are named after Dr Percy Cliffe and his wife, Dr Carice Ellison, who endowed the lecture to be given on a subject connected with the contribution of fundamental science to the advancement of medicine. The Lecturer is also awarded a medal in honour of their presentation. Retrieved from https://en.wikipedia.org/wiki/Ellison%E2%80%93Cliffe_Lecture.

[iv] *Conversations with Pathologists* Pathological Society of Great Britain 25th July 2007. Transcript of interview with Julia Polak at her home. Retrieved from http://www.pathsoc.org/conversations/index.php?view=article&catId=60%3Ajulia-polak&id=80%3Ajulia-polak-full-transcript&option=com_content

[v] *Ibid.*, iv *Conversations*, Section 1, p. 3.

[vi] *Op cit.*, iv *Conversations*, p. 17.

[vii] *Op cit.*, iv *Conversations*, Section 4, p. 14.

[viii] *Op cit.*, iv *Conversations*, Profile.

[ix] Women in Cell Science by Fiona Watt, Editor-in-Chief JCS, *J. Cell. Sci.* 117, pp. 5195–5196.

[x] *Op cit.*, iv *Conversations*, p. 2.

[xi] Moises Polak

[xii] Time of the disappeared

[xiii] Professor Antony Guy Everson Pearse (1916–2003) an innovator, he developed histopathology for research and diagnosis the first Professor of

Histochemistry 1965 Munk's Roll XI 1998–2004 Royal College of Physicians Lives of the Fellows. Retrieved from http://munksroll.rcplondon.ac.uk/Biography/Details/5358.

[iv] *Op cit.*, iv *Conversations*, Section 1, p. 2.

[xv] *Op cit.*, iv *Conversations*, p. 3.

[xvi] *Op cit.*, iv *Conversations*, Section 1, p. 2.

[xvii] Professor Sir Stephen Bloom FRS (1942–) Imperial College London: Steve Bloom is the Head of Division for Diabetes, Endocrinology and Metabolism, Chair of the academic Section of Endocrinology and Investigative Medicine at Imperial College London and Lead Clinician for Clinical Chemistry at Imperial College Healthcare NHS Trust. He heads a 40 strong research team investigating the physiology of regulatory peptides in CNS and periphery. Retrieved from http://www.imperial.ac.uk/people/s.bloom. Accessed 25 April 2016.

[xviii] Sir Magdi Habib Yacoub OBE FRS (1935–) Professor of Cardiothoracic Surgery at the National Heart and Lung Institute, Imperial College London and Founder and Director of Research at the Harefield Heart Science Centre (Magdi Yacoub Institute) overseeing over 60 scientists and students in the areas of tissue engineering, myocardial regeneration, stem cell biology, end stage heart failure and transplant immunology. He is also Founder and Director of Magdi Yacoub Research Network which has created the Qatar Cardiovascular Research Center in collaboration with Qatar Foundation and Hamad Medical Corporation. Retrieved from http://www.imperial.ac.uk/people/m.yacoub. Accessed 25 April 2016.

[xix] *Op cit.*, iv *Conversations*, p. 11.

[xx] *Ibid.*, iv *Conversations*, p. 11.

[xxi] *Op cit.*, iv *Conversations*, Section 4, p. 14.

[xxii] *Op cit.*, iv *Conversations*, Profile http://en.wikipedia.org/wiki/Argentina

[xxiii] *Op cit.*, iv *Conversations,* p. 70

[xxiv] Retrieved from http://www1.imperial.ac.uk/departmentofmedicine/divisions/experimentalmedicine/pathology/jprt/

[xxv] *Op cit.*, iv *Conversations*, Section 2, p. 7.

[xxvi] *Op cit.*, iv *Conversations*, Section 4, p. 14.

[xxvii] *Op cit.*, iv *Conversations*, p. 16.

[xxviii] Professor Larry Hench (1945–2015) Emeritus Professor of Ceramic Materials Imperial College London 1960s engineered Bioglass, a material that could be used to assist in bone repair, developed this and other systems with Julia Polak. Emeritus Co-Director Centre for Tissue Engineering and Regenerative Medicine Imperial College London.

[xxix] *Op cit.*, iv *Conversations*, p. 17.

[xxx] *Ibid.*, iv *Conversations*, p. 17.

[xxxi] *Ibid.*, iv *Conversations*.

[xxxii] *Ibid.*, iv *Conversations*, p. 20.

[xxxiii] *Ibid.*, iv *Conversations*, p. 20.

[xxxiv] "The Imperial College pledge to our academic women, is that we shall take all necessary action to create and maintain a level playing field …" Retrieved from http://www3.imperial.ac.uk/equality/staffnetworksandcommittees/aoc/pledge.

[xxxv] *Op cit.*, iv *Conversations*, p. 22.

[xxxvi] *Op cit.*, i *Conversations*, p. 12.

Medicine

Haematology

Dame Janet Maria Vaughan

Name and Post Nominals

Dame Janet Maria Vaughan (1899–1993)
DBE, OBE, FRS, MRCP, FRCP, DM, B.M, B.Ch.

First woman Councillor, Royal College of Physicians.

Specialism

Haematology; Anaemias; Pathology; Blood transfusion and blood transfusion service; Bone, and marrow pathology and radiological effects; medical education.

Career Roles

Pathologist RPMS 1938; Beit Memorial Fellowship and Leverhulme Fellowships for Research London Hospital Medical School; Assistant Clinical Pathologist, RPMS 1934; Mistress, Somerville College Oxford 1945–1967.

Career Awards

MRCP 1933; FRCP 1939; FRS 1979; Osler Medal Oxford University 1970; Fogarty Scholar National Institutes of Health Bethesda 1973;

Honorary Degrees: DSc. University of Wales 1964; DCL Oxford 1967; DCL London 1968; DSc Leeds 1970; DCL Bristol 1971; DSc Liverpool 1973. Women of Our Century Interview BBC Two 1984.

Awards

OBE 1944; DBE 1957.

Professional Affiliations

Member, Goodenough Committee on Medical Schools 1942; Trustee, Nuffield Foundation Funding 1943–1967; Member, Royal Commission on Equal Pay 1944–1946; Vice Chairman, Oxford Regional Hospital Board 1948, Chairman, 1950–1951; Member, Board of Governors Oxford University Teaching Hospital 1950–1951; Phillips Committee on the Economic and Financial Problems of the Provision for Old Age 1953–1954; Medical Advisory Committee University Grants Committee, Member, Committee on Libraries 1953–1963; International Committee for Radiological Protection Task Groups 1967 and 1969.

Lectures

Quote

> To be remembered as a scientist. That I have been able to solve, to throw light onto fascinating problems. But as a scientist who had a family. I don't want to be thought of as a scientist who just sat thinking. It's important you have a human life.

Born

18th October 1899 Clifton Bristol.

Died

9th January 1993.

Family Background

Father: William Wyamar Vaughan (1865–1938), Assistant Master at Clifton College Bristol, later headmaster Giggleswick School; Master Wellington College, Headmaster Rugby school; Mother: Margaret Symonds (1869–1925) daughter of the historian J.A. Symonds. Siblings: three, one sister who died in childhood, two brothers.

Marriage

1930 David Gourlay (1889–1963), Civil Servant — Post Office, later founder of the Wayfarers' Travel Agency. Two daughters, Mary b. 1933 and Priscilla b. 1935.

Education

Schooling

At home; North Foreland Lodge.

University

Undergraduate

Somerville College Oxford 1919–1922.

Degree

BM BCh.

Postgraduate

Oxford University.

Degree

MD.

Postgraduate Awards/scholarships

Rockefeller Foundation Travelling Fellowship 1929–1930; Beit Memorial Fellowship 1930–1933; Leverhulme Fellowship 1933–1934.

Text Discussing Career and Other Comments

Janet wanted to make a difference in life. Her thoughts turned to medicine as a career, in order to make a difference in society. This she achieved, through pathology, haematology, blood transfusion and organisation of services, effects of nuclear fission on bone and medical education. Medics were part of her extended family going back 300 years; her great–great uncle had served four monarchs from George III to the young Victoria. Although she was taught no science, she did understand the value of a medical qualification to give her that standing in society she wanted in order to make a difference. Throughout her life she was very concerned with social injustice, the social implications of medicine and the need to relieve poverty and lowly living conditions.

Her ambition was nearly thwarted by the headmistress at her school — when she eventually attended one that is, (North Foreland Lodge) from the age of 16, who said she was "too stupid to be worth educating."[i] There was clearly no substance to this statement, but having been educated by governesses up to then, her breadth and depth of knowledge and intelligence had perhaps not been best developed in a formal method of education.

Janet was not interested in languages, but in most other subjects, and displayed leanings towards the left politically, and to social politics. A measure of her determination to be in a position to influence society was that she made three attempts to pass the exams for entry to Somerville, and on the third did so. Although she had no science background at all (botanical drawing from life but nothing else), her intention, when she went up in 1919, was to become a doctor. She took a Physiology degree and from a standing start of no knowledge, by dint of application and reading — despite as she says, being a danger in the lab because of this,[ii] graduated first class. Her tutors included H.C. Bassett, whose reading recommendation of a classic textbook of physiology, widened her medical horizons for life. J.B.S. Haldane[iii] and Charles Sherrington,[iv] also tutors were both very supportive of the young woman student in the very male bastion of Oxford. She needed a male chaperone to attend meetings of the political group the Mazzini Society, although elected to it. Attendance may have been influenced by Haldane and her fellow student Margaret Huxley, as she socialised with their families. Margaret was the sister of Julian Huxley, who was a Fellow at new College and Aldous Huxley the writer.[v] Janet was President of Somerville Debating Society when Vera Brittain, another fellow student, proposed that travel was superior to university life, which then appears in Vera's 1933 memoir of the futility of WWI, *Testament of Youth.*

To fulfil her ambition to become a doctor, she needed clinical training which she obtained at University College Hospital (UCH), completing this in 1925, graduating B.M.B.Ch. She was much affected by the state of the health of the general population she came across in her training; the awful dirty and almost inhumane conditions the poor had to live in. Their lack of affordable good food and nutrition. This inequality that she wished to rectify remained with her throughout her life. Showing the signs of flexibility and pragmatism that informed her later work, Janet's direction changed after qualifying, she turned to training as clinical pathologist in order to have free time to undertake family duties.

This is when her interest in haematology emerged, beginning to understand the impact of poor nutrition on health, and in particular, anaemia and the debilitating effects of this illness. She undertook ward rounds in order to get to know her consultants patients. In doing this with the rest of the medical team, she came across medical statistics for the first time, and realised how important they are, even though this was a relatively new field at that time.[vi] She saw the value of knowing the size of red blood cells in anaemia for example.

Always well read, she discovered that an American physician, George Minot, gave liver to his patients who were suffering from pernicious anaemia.[vii] The results looked good, so supported by the chemist Sir Charles Harington,[viii] she determined to try this. She recruited friends to try her concoctions of minced liver, including her relation Virginia Woolf and other Bloomsbury-ites. In fact Virginia Woolf refers to the experiment in her book *A Room of One's Own*.[ix] The results tested on dogs were not a success, but Janet took it to no ill effect, and she was allowed to try it with patients, to great positive effect. (That an available treatment used by others, Fowlers Solution, containing poisonous arsenical compounds didn't work, was not surprising.) Janet's dietary system must have resulted not only in more lives being saved, but also remission of the disease. This experimental work formed the basis of her 1931 DM thesis.

On becoming free of family restriction's, in 1929, Janet decided to study in America at Harvard at the Thorndike Memorial Laboratory Boston City Hospital, run by George Minot (later Nobel Prize winner for his work)[x] whose research she had instigated at UCH; she was granted a Rockefeller Travelling Fellowship to fund her travel. (She also visited other labs whilst in America.) A friendship developed between Janet and Dr and Mrs Minot, but nonetheless he could not help the rule that forbade woman from working with patients. In the end, she worked with pigeons, very successfully. Because of their role they were known as the "bloody pigeons". In fact, they proved the value of vitamin B12 in pernicious anaemia, in the first ever test for it. But it wasn't until 50 years later that the scientist she was working with, Bill Castle,[xi] gave her the credit for the experiment in a published paper. In 1930, although she had been asked to stay on for another year, given the circumstances of the mores she had to work under, she took the best route for her — which was back to England and to marry her fiancé, David Gourlay. They had met whilst she was at UCL and lived in a house shared with others associated with the War Victims Relief Organisation in France.

Janet and David saw her career as important and there was no question of her giving up her work. They shared strong views on social injustice and poverty and she supported David in giving up his civil service career and founding the Wayfarers Travel Agency. This innovative company included educational school trips (not something that was generally undertaken at that period) and was intended to broaden international relations and understanding from an early age, and thus prevent war. David is described as a socialist and internationalist in a list of WWI conscientious objectors.[xii] His stance emphasises the strength of feeling he and Janet shared. He taught during WWI, and using education, followed up his convictions by developing the scheme for early life development of international collaboration.

Following her time in the US, funded by a Beit Memorial Fellowship and a Leverhulme Fellowship, Janet developed her study of blood cell formation (haematopoiesis) and the relationship of bone marrow whilst working at The London Hospital, but found the male staff generally unwelcoming and unhelpful.

In 1933, her first daughter was born, and she carried on her researches, supported, as always in her working life, by David. In 1934, she was invited to work at the British (later Royal) Postgraduate Medical School, at the Hammersmith Hospital, by E.H. Kettle,[xiii] who was Head of the Pathology Department. Her title was Assistant in Clinical Pathology, which meant that she undertook routine work, including blood transfusions, but Janet insisted that she should continue her research and gradually built up a team in that area. She also published what was to become the standard textbook for the subject: *The Anaemias.*[xiv]

In 1935, her second daughter was born, and Janet continued working.

The following year, the Spanish Civil War began, and medical responses sparked off a new direction in Janet's career: blood transfusion services. Involved in the Committee for Spanish Medical Aid during the Spanish Civil War, she had realised the need for blood supply for the wounded. She began work on this in a small way at the Hammersmith, and with like-minded people in London medicine, she led and they prepared for the creation of the National Blood Transfusion Service in 1938, submitting their plan to the MRC unsolicited. Janet's boss called her "pretty naughty"[xv] for doing this — how wrong could he be! The work encompassed the transition from peacetime to wartime requirements — she feared the onset of a second world war. Everything needed to be developed and scaled-up, from the very basic invention of equipment — blood collection containers, needle types, to the new problem of large scale blood storage and techniques for treatments, from peacetime injuries to wartime injuries, for large scale bomb and burns victims, to the needs of the fighting troops and civilian populations. This was not perhaps what she had in mind for her career, but it was a result of the need she foresaw in preparation for wartime. Experts in haematology were needed and she displayed all necessary abilities: knowledge, ingenuity, calm organisational skills and good communication skills.

In setting up her system, she began at grassroots level with £100 and purchase from ordinary shops of glassware and rubber tubing as she realised there would be a great deal of 'bleeding' (taking blood) needed. There were long discussions on suitable containers — a milk bottle style was settled and known as the "Janet Vaughan" or the "MRC" . As preparations for war went on, the MRC took over the transfusion organisation. Blood transfusion stations were organised. She set up and ran, the North West London Blood Supply Depot, Slough, as Medical Officer in Charge. Janet and her teams not only took blood (called bleeding then) from volunteers and matched it for type, and worked out how to store it — giant fridges were built at the centres — ice cream vans commandeered for refrigerated transport, but they also went out to deal with mass casualties during WWII. From this they learnt what equipment was needed, how it needed to be adapted to situations and how blood could be got in to people with various types of injuries, including through bone.

Understanding nutrition and how it worked with the complex bodily system of blood creation and health had been a major part of her career. Postwar therefore, Janet was asked to attend liberated European cities, and eventually the Belsen Concentration Camp, to help treat starving people. An experimental protein fluid was provided, given intravenously or by mouth. These experiences were quite horrifying, but Janet remained calm and professional and quickly gauged what was needed, that is small amounts of ordinary food and milk. On returning she informed the authorities that the fluids provided were useless; production of the protein fluids was stopped.

She had other victories, such as the Royal College of Physicians asking her to work with colleagues to report on the need for engagement with social medicine, which vindicated her insistence on this need over some years — this was prior to the National Health Service being created.

At the end of the war, the use of the atomic bomb in Japan worried her, for the impact on the health of survivors. She knew that bones would be affected and set up research experiments with various radioactive substances in order to understand the effects of them. The MRC funded the team and Janet's output on effects of plutonium on bone led to global recognition. (She published three books on bone in the 1970s in her retirement.) In 1945 though, Janet continued her research even as she became

Mistress of Somerville College. She was also glad to move her family to Oxford, a good place to bring up children she and David thought, especially as they had been bombed out of their London home. She was known to complete her Somerville work early so that she could spend the day in her lab. She continued researching into her 80s. One of her students at Somerville was Margaret Roberts, later Baroness Thatcher the first woman UK Prime Minister, who graduated with a chemistry degree in 1947. Janet made a huge impact on life at Somerville and the University of Oxford, advocating for science teaching and mixed gender colleges.

Janet's career was highly successful, in the face of opposition and obstruction from men, who treated her like a child at times, perhaps balanced out by those who did support her, with whom she continued her research. Janet's work was ground breaking, innovative in materials and methods, imaginative and of great value to society — she created her own sphere of social medicine.

Endnotes

[i] Dame Janet Vaughan in interview with Max Blythe 19871104 Royal College of Physicians and Oxford Brookes University Wellcome Video Sciences Archive MSVA 027.

[ii] *Ibid.*

[iii] John Burdon Sanderson Haldane (1892–1964) Physiologist, Evolutionary Biologist, Biostatistician, Geneticist, Mathematician.

[iv] Charles Scott Sherrington (1857–1952) Physiologist.

[v] Margaret (1889–1981), Julian (1887–1975) and Aldous (1894–1963) Huxley were grandchildren of Thomas Henry Huxley, (1825–1895) Professor at the Royal School of Mines, biologist known as Darwin's Bulldog for supporting Charles Darwin's Theory of Evolution. Huxley family. Retrieved from https://en.wikipedia.org/wiki/Huxley_family.

[vi] Cecil Price-Jones (1863–1943) Pathologist Developed system for measuring red blood cells: The Price-Jones Curve.

[vii] Anaemia caused by lack of vitamin B12 resulting in a low red blood cell count.

[viii] Sir Charles Harington FRS (1897–1972) Professor of Chemical Pathology University College Hospital 1931–1942; and Director National Institute for medical research 1942–1962. Retrieved from https://en.wikipedia.org/wiki/Charles_Robert_Harington. Accessed 27 May 2016.

[ix] Virginia Woolf (1882–1941) Writer. *A Room of One's Own* 1929.

[x] George Richards Minot (1885–1950) Joint Nobel Prize for Physiology or Medicine with George Hoyt Whipple and William P. Murphy for work on pernicious anaemia 1934. Retrieved from https://en.wikipedia.org/wiki/George_Minot. Accessed 27 May 2016.

[xi] William Bosworth Castle (1897–1990) American physician, physiologist, haematologist. Retrieved from https://en.wikipedia.org/wiki/William_Bosworth_Castle. Accessed 27 May 2016.

[xii] Surrey Men in the Pearce Register of Conscientious Objectors. Retrieved from www.surreyinthegreatwar.org.uk/. Accessed 31 May 2016.

[xiii] Edgar Hartley Kettle (1882–1936) Pathologist. Retrieved from http://munksroll.rcplondon.ac.uk/Biography/Details/2552. Accessed 27 May 2016.

[xiv] The Anaemias Oxford University Press 1934.

[xv] How did Blood Transfusion Work in Britain in WWII and its significance *Journal of Archaeology, History and Anthropology* 21. 2008 Sreedha Krishna, Student Faculty of medicine Imperial College. Retrieved from www.scientific-journals.org/journals2008/articles/1395.pdf. Accessed 28 June 2016.

Source: Some Women Pioneers In Haematology Barry G. Firkin Version of Record online: 24, December 2001, *British Journal of Haematology* **108**(1), pp. 6–12, 2000.

Richard Doll, 'Vaughan, Dame Janet Maria (1899–1993)', *Oxford Dictionary of National Biography*, Oxford University Press, 2004; online edn, May 2010. Retrieved from http://www.oxforddnb.com.ezproxy2.londonlibrary.co.uk/view/article/42277. Accessed 27 May 2016 Dame Janet Maria Vaughan (1899–1993).

Dame Janet Maria Vaughan *Biographical Memoirs of the Fellows of the Royal Society* Maureen Owen **41**, pp. 485–498, 1995.

Winifred May Watkins

Name and Post Nominals

Winifred May Watkins (1924–2003)
FRS, PhD, FMEdSci, BSc.

Specialism

Haematology: Relationships in Blood Group Systems, Blood Transfusion; Blood Disorders — leukaemia; Complex Carbohydrates.

Career Roles

Laboratory Assistant, Lister Institute 1942; Sabbatical year University of California Berkeley lab of William Z. Hassid 1960; Reader in Biochemistry 1965, Professor of Biochemistry and Head of the Biochemistry Department Lister Institute University of London 1968–1975; Head of Immunochemical Genetics MRC Northwick Park 1975–1989; Honorary Member Haematology Department Hammersmith Hospital 1989.

Career Awards

Oliver Memorial Fund Annual Award 1965; Jointly with Walter Morgan the Landsteiner Memorial Award American Association of Blood Banks

1967; Jointly the Ehrlich–Ludwig Darmstädter Prize 1969; Fellow, 1969; Member of Council, 1984–1986, Royal Medal 1988 Royal Society; William Julius Mickle Fellowship University of London 1970; Kenneth Goldsmith Award of the British Blood Transfusion Society 1986; Founder Member and Fellow of the Academy of Medical Sciences 1998; Franz Oehlecker Medal, German Society of Transfusion Medicine and Immunohaematology 1989; Jointly with Walter Morgan the Philip Levene Award of the American Society of Chemical Pathologists 1990; Honorary DSc Utrecht 1990.

Awards

Foreign membership of: The Polish Academy of Science and the Royal Swedish Academy of Science; Honorary Fellowship Royal College of Physicians; Honorary Member International Blood Transfusion Society; Honorary Member Japanese Biochemical Society.

Professional Affiliations

Royal Society; Academy of Medical Sciences; Foreign Member Polish Academy of Science 1986.

Quote

> "To be a guest worker in [Winifred's] lab be it for a year or an afternoon, was to enjoy the work and to feel a sense of excitement."[i]

Born

6th August 1924 Shepherds Bush, London.

Died

3rd October 2003 Stanwell, Surrey.

Family Background

Father: Albert Edward Watkins an engraver with art as a hobby; Mother: Annie Bertha nėe Delia, a housewife; Siblings: one sister.

Marriage

None.

Education

Schooling

St Stephen's Parish Elementary School Shepherds Bush, Godolphin and Latymer Girls School Hammersmith.

Scholarships

London County Council for Secondary School, Chelsea Polytechnic University London (external).

University

Undergraduate

Chelsea Polytechnic. University London (external).

Degree

BSc Hons.

Postgraduate

St Bartholomew's Hospital Medical School.

Degree

PhD.

Text Discussing Career and Other Comments

Winifred May Watkins was born into an unremarkable family in Shepherds Bush, west London. Her father was a process engraver and an artist in his spare time; her mother was a housewife, a not unusual and most acceptable role at that time.

Winifred's early education was at her local church school, St Stephen's Parish Elementary. Her secondary school was the prestigious Godolphin and Latymer Girls School in Hammersmith, to which she won a London County Council Scholarship. Not far from home, this school was within easy walking distance of her home. That is, until the outbreak of World War II when the school was peremptorily removed to Windsor and then to Newbery, but this was to be a place where Winifred found her vocation. She had access to the science labs at Newbery Boys School and so her interest and talents for science were kindled. Despite their move, the integrity of the school and its ethos was retained. Winifred was instilled with positive values, self-discipline, confidence in her abilities, awareness of others, both immediately around her and internationally and personal responsibility. This was a happy time for her and she kept in touch with the school throughout her life. The values she honed there stood her in good stead for her later life.

The school returned to London, and in 1942 she passed her Higher School Certificate. Her grades would have won her a university place, but because of the war, and strictures on university access, it was only a job, not higher education or a career that was open to her. She made the best of this however, taking up work as a Laboratory Assistant at the Lister Institute of Preventative Medicine in Chelsea London, (now the Lister Hospital site) so that she could keep up her work in science. Medical research was a reserved occupation as it was essential to the war effort, so this meant Winifred would not be called up into any other service. The work she was engaged in was particularly relevant to war as it concerned blood and blood transfusion, an area that required much more understanding to enable safe use.

The war had pros and cons for young people — prevented some levels of education, but broke down other barriers. At the Lister, Winifred had access to many senior scientists that in more usual times she would not have had any contact with. It was a strange time, as in so many organisations, parts remained in London and other parts went elsewhere. It was only the Biochemistry Department of the Lister that remained, but this made it all the more cohesive and intimate. She was party to intellectual discussions debate and scientific processes. Winifred first worked with the eminent haematologist Walter Morgan, who had been investigating blood group antigens[ii] since 1940. Very quick to learn, and to produce data, she was able to write a paper in a very short time, but not all barriers had been broken down. She was only a lab assistant, was young and unqualified, this was not the usual sort of background from which professional papers came. She had to get the permission of the Director to publish, but her name did appear on the paper published in 1944. By 1952 her name was appearing on papers as lead author and investigator.

The collaboration between Winifred and Morgan was to be life long as colleagues on outstanding research in haematology, and as friends. She helped to look after him in his old age after his wife died.

Drawing on her innate abilities and those values inculcated by her school, Winifred felt encouraged to take up courses at the nearby Chelsea Polytechnic in the evenings. This led to her BSc Hons in Chemistry, ratified by the University of London, in 1947. It is not easy to study by day and work at night, particularly under unpredictable war time conditions of air raids, difficult transport and rationing. (Although the war ended in 1945, rationing continued to be applied to various items until 1954. There also remained a lot of clearing up to do following bombing raids, collapsed buildings and societal infrastructure to accommodate a new way of life for everyone, reintegrating those who had been in the forces serving away from home.)

On gaining her degree, Winifred started full time postgraduate work at St Bartholomew's Hospital Medical School, though she returned to assist at the Lister during evenings and weekends! A testament to her enthusiasm and gift for work. The immunochemist, Arthur Wormall was her supervisor. Her work involved studying compounds used to treat leukaemia and their impact on the immunology of the body. This was quite new area of investigation. She published papers during her studentship and was awarded her PhD in 1950.

Winifred then returned to the Lister, now eminently qualified to work in her own right. She joined Walter Morgan's Blood Group Research Unit. She made major discoveries about the relationships of the blood groups which had a huge impact on the area of blood transfusions and their safe use. Her work wasn't always received positively, but her values enabled her to rise above criticism until her research came to be accepted and proved right. The work she and Morgan undertook has been noted as all the more remarkable because it was molecular biology at a time before many helpful tools had been devised.[iii]

In 1960, she left for a years' work at the University of California at Berkeley on a Henry Wellcome Travel Fellowship, here she made further discoveries in relation to blood genes and enzymes, and in 1965 was given The Oliver Memorial Fund Annual Award for her superb work on blood transfusion. This was followed by many awards and in 1969, she was elected a Fellow of the Royal Society and became a member of its council between 1984 and 1986. She was awarded the Royal Society's Royal Medal in 1988.

Winifred's work continued at The Lister, with a Professorship from 1968 (only the Lister's 4th), until its closure in 1975,[iv] when she transferred her lab to an MRC facility at Northwick Park, appointed Head of the Immunochemical Genetics section. When this closed, she transferred to the Hammersmith Hospital Haematology Department in 1989 and retired from there in 2000.

Of course Professors rarely retire fully, and indeed why should they when they have so much more to contribute? Winifred continued her research but had already suffered a stroke in the 1970s which did not impede her work, but following further strokes in 2003 subsequently died.

Winifred believed both sexes should have the best educational opportunities available to them and was a governor of two schools — Dulwich College and Alleyns School between 1987 and 1995 where she could put her principles into practice.

Although good natured and good fun, Winifred did not marry, she had a close immediate family and three nieces, and for most of her career did not stray far from her original home of Shepherds Bush, but she made a global impact on haematology and immunochemistry.

Endnotes

[i] Robin Marshall a long term colleague of Winifred's in Winifred May Watkins Obituary, pp. 56–59. *The Biochemist*, June 2004. The Biochemistry Society. retrieved from https://www.jstage.jst.go.jp/article/tigg1989/11/62/11_62_391/_pdf.

[ii] ODNB P. L. Mollison, 'Watkins, Winifred May (1924–2003)', *Oxford Dictionary of National Biography*, Oxford University Press, Jan 2007; online edn, January 2009.

[iii] *Ibid.*

[iv] Now a science funding body.

Sources: ODNB P. L. Mollison, 'Watkins, Winifred May (1924–2003)', *Oxford Dictionary of National Biography*, Oxford University Press, Jan 2007; online edn, January 2009.

Cardiology

Celia Mary Oakley

Name and Post Nominals

Celia Mary Oakley (Pridie) (1931–2014), MRCS LRCP, MBBS, MRCP, MD, FRCP, FACC, FESC.

First woman to be appointed House Surgeon to the celebrated cardiologist Paul Wood 1957.

One of only three women cardiologists in the UK in 1958.

Specialism

Cardiology.

Career Roles

House physician and house surgeon at the Royal Free 1953–1957; House physician to cardiologist Paul Wood Brompton Hospital 1958; National Hospital, Queen Square; Postgraduate Medical School at Hammersmith Hospital here, part of the team that coined the term 'hypertrophic obstructive cardiomyopathy'; Professor of Clinical Cardiology Hammersmith Hospital 1991; Private Practice; Medico–legal consultant.

Career Awards

Medical Research Council Fellowship Cardiopulmonary Laboratory University of Rochester, New York 1963; FRCP Fellow Royal College Physicians 1970; FACC Fellow American College Cardiology 1972; 1970 FESC Fellow European Society of Cardiology 1970; Personal Chair in Cardiology 1991; Laennec Master Clinician Award from the American Heart Association 2004; Mackenzie Medal of the British Cardiovascular Society for services to British cardiology 2006.

Professional Affiliations

Committee on Safety of Medicines. Founding Fellow of the European Society of Cardiology, Chairman of the Society's working group on valve disease 1995–1998 and of the task force on the management of heart disease in pregnancy published guidelines in 2003.

Lectures

Given by Her

Scottish Society of Physicians Fitzgerald Peel Lectures, 1986. *Non Ischaemic Disorders of the Heart Muscle.*

Quote

> Celia Oakley the doyen of London cardiology.[i]

Born

14 May 1931 Bishop's Castle, Shropshire.

Died

17 November 2014.

Family Background

Father: Arthur Howard Oakley, Bank Manager; Mother: Minnie Isabel Oakley née Stevenson.

Marriage

1950 Ron Pridie Pioneer Echocardiographer, Harefield Hospital; two daughters, four grandchildren.

Education

Schooling

Berkhamsted School.

University

Undergraduate

Royal Free Hospital School of Medicine 1954.

Degrees

MRCS LRCP 1954 MBBS London 1954.

Undergraduate Awards

Scholarships and five prizes.

Postgraduate

University of London.

Degrees

MRCP 1956; MD 1965; FRCP 1970; FACC 1972 FESC.

Text Discussing Career and Other Comments

Celia developed her career against the two obstacles of male dominance and outright opposition she encountered. She got her first house job with Paul Wood at the Brompton Heart Hospital, by having the courage to ask him if she applied would she be considered. She had been told she would not be recommended for this plum job as it shouldn't go to a woman, but she did apply and got it!

Later in her career, a move to another hospital for a consultant post was blocked by a consultant refusing to work with a woman. This may have deterred others but Celia seemed to take it in her stride. She remained at Hammersmith Hospital where her career went from strength to strength.

She demonstrated the confidence she had in her abilities, which others shared, in her superb contribution to treatment of patients and research — she published over 400 papers, a quarter of these as sole author. Celia was a highly valued member of the cardiology team at the Hammersmith. It was at the Hammersmith that pioneering work was undertaken in treatment of heart disease in the 1950s and 1960s.

As surgical treatments developed, Celia became interested in the management of congenital conditions that would previously have been terminal much sooner.

Celia became known for her innovative approach to cardiology and had five years of funding from the British Heart Foundation for work on children's heart problems — until non-paediatricians were prevented from working with child cardiac cases.

Endnotes

[i] Mervyn Gotsman MD. Retrieved from http://mervyngotsman.blogspot.co.uk/2013/05/london-london-had-been-theepicenter-of.html. Accessed 3 July 2016.

Sources: Munk's *Roll Lives of the Fellows* Volume XXII 2005 — Royal College of Physicians Historic Moments in Cardiology:Celia Oakley, MD, FRCP, FACC, FESC *Circulation* 26, January 2010.

Dame Margaret Elizabeth Harvey Turner-Warwick

Name and Post Nominals

Margaret Elizabeth Harvey
Turner-Warwick (1924–)
DBE, DM, PhD, FRCP, FFOM, FRACP, FRCPE, FFPHM, FRCP, FRCGP, FRCPI.

First Woman President Royal College of Physicians 1989–1992.

Specialism

Thoracic medicine.

Career Roles

Postgraduate training University College Hospital and Royal Brompton Hospital; Lecturer, Senior Lecturer, Diseases of the Chest Cardiothoracic Hospital London; Consultant Physician, Elizabeth Garrett Anderson Hospital Royal Brompton and London Chest Hospitals 1965; Professor of Medicine (Thoracic Medicine), 1972–1987, Dean, 1984–1987, Cardiothoracic Institute, then Emeritus Professor.

Career Awards

Honorary DSc's from Universities of New York 1985; London 1990; Exeter 1990; Hull 1991; Oxford 1992; Sussex 1992; Cambridge 1993; Leicester 1998; Fellow of the UK Academy of Medical Sciences 1989; Honorary Fellow Lady Margaret Hall University of Oxford 1991; Fellow University College London 1991; Osler Memorial Medal at Oxford 1995; Fellow Imperial College 1996; Honorary Fellow Green Templeton College Oxford; Margaret Turner-Warwick Library; President's Award European Respiratory Society 1997; Presidents Medal British Thoracic Society 1999. Margaret Turner-Warwick Education Centre for the NHLI at the Royal Brompton Campus 2016. DM Oxon, 1956; PhD London 1961; FRCP 1969; FFOM 1983; FRACP 1983; FRCPE 1990; FFPHM 1990; FRCP Glasgow 1991; FRCGP 1991; FRCPI 1992.

Awards

DBE 1991; The Honourable Society of the Middle Temple Honorary Bencher 1989.

Professional Affiliations

President British Thoracic Society 1982–1985; Medical Research Council Systems Board (DHSS nomination), 1982–1985; Chairman Central Academic Council, BPMF, 1982–1985; Asthma Research Council and Chairman Medical Research Committee 1982–1987; Council 1984–1990, Vice President British Lung Foundation 2008–2009; Royal College of Physicians Second Vice President 1988–1989, President 1989–1992; UK Co-ordinating Committee on Cancer Research 1991–1997; Chairman, Royal Devon and Exeter Health Care NHS Trust, 1992–1995 King's Fund General Council 1991; Council, British Heart Foundation Council 1994–; ASH Council and Vice President 1990; University of London: Senate 1983–1987; Academic Council, 1983–1987; Scholarships Committee 1984–1987; Committee Extramural Studies, 1984–1987; Founder member of the UK Academy of Medical Sciences 1989; Academy of Royal Medical Colleges, previously known as: Conference of Colleges and their Faculties in UK 1990–1992; Nuffield Bioethics Council 1993–; Round Table on Sustainable Development 1995–1998.

Lectures

Given by Her

1974 Mark Daniel, 1980 Philip Ellman, 1985 Tudor Edwards, 1994 Harveian Oration Royal College of Physicians 1994[i]; 1982 Lettsomian Lectures Medical Society of London.

Lectures

In Her Name

The Annual Margaret Turner-Warwick Respiratory Lecture NHLI and Royal Brompton Hospital.[ii]

Quote

> Systematically ignore gender prejudice because it is irrelevant to medicine and other management situations.[iii]

Born

19th November 1924.

Family Background

Father: William Harvey Moore QC; Mother: Maud Baden-Powell; Siblings: three sisters.

Marriage

1950 Richard Trevor Turner-Warwick (1925–) CBE 1998, Surgeon (Hunterian Professor Royal College of Surgeons) Distinguished Urologist.[iv] Family: two daughters, six grandchildren, two great grandchildren (possibly four by the time of publication, Dame Margaret writes).

Education

Schooling

City of London School for Girls; Walthamstow Hall Sevenoaks Kent; Maynard School Exeter; St Paul's Girl's School London.

University

Undergraduate

Physiology Lady Margaret Hall, Oxford University 1943–1946.

Degree

DM Oxon 1946.

Undergraduate awards/scholarships

Open Scholarship.

Postgraduate

University College and Hospital London, 1947–1950.

Degree

PhD 1961.

Post Graduate awards/scholarships

University College Hospital London 1947–1950 Student prizes: Tuke Silver Medal, Filliter Exhibition in Pathology, Magrath Scholarship in Medicine, Atchison Scholarship for outstanding student of the year, University College Hospital.

Text Discussing Career and Other Comments

Dame Margaret has had first hand experience of the need for thoracic medicine as she was diagnosed with tuberculosis in the final year of her first degree, and after graduating spent a year in a Swiss sanatorium, rest and fresh air providing much of the treatment before the development of antibiotics. Declared fit for work in 1947, she became a post graduate at University College Hospital London, and the Royal Brompton Hospital London, where she gained awards for her work as a student, and further qualifications including in 1961, her PhD, which interestingly was based on her clinical work which she undertook as she felt it needed to be done. This was perfoming autopsies (with all necessary permissions and support from the hospital authorities) on patients who had not responded to treatment to look at their lungs to discern and investigate findings to improve treatments.

She subsequently gained qualifications from her professional body, the Royal College of Physicians. She studied for the membership exams whilst on three months maternity leave. This she says was an advantage not then open to her male colleagues.[v]

In 1961, she took a post as Lecturer, then Senior Lecturer in Diseases of the Chest at the Cardiothoracic Hospital and also became Consultant Physician at the Elizabeth Garrett Anderson Hospital. She left this post in 1967 for that of Consultant at the Royal Brompton and London Chest Hospitals, becoming a Fellow of the Royal College of Physicians in 1969. In 1972, Dame Margaret became Professor of Medicine, a position she retained until her retirement in 1987, when she became Professor Emeritus, although she continued postgraduate teaching. She had gained her Fellowship of the Faculty of Occupational Health of the Royal College of Physicians in 1983. Simultaneously to her Consultant's post and Professorship, she was Dean of the Cardiothoracic Institute between 1984 and 1987, during which time she lead the team to fund, convert and relocate the Cardiothoracic Institute and was President of the British Thoracic Society between 1982 and 1985.

In 1988, she was voted Second Vice President of the Royal College of Physicians, a post not usually expected to lead to the Presidency, however in Dame Margaret's case, it can be said to have done so. She was elected President in 1989. This was a very prestigious and well earned position that she held until 1992. She was their first woman President (or lady President) as she says, in their 472 year history![vi]

The election is tough and held in a special meeting of the College of Physicians. The election to the Presidency speaks volumes for the impression Dame Margaret has made on her profession and colleagues. Her modesty, impeccable manners, attention to her work and devotion to her patients and to the development of her profession had been rewarded. Not that she would have expected any reward, at this point she was intending to wind down to retirement.

Her many other appointments and involvements in medical and educational interests are listed at the beginning of the biography. Dame Margaret's interests were in postgraduate teaching of international students and collaborative teaching with overseas professors on sabbatical on extended weekly clinical ward rounds, as well as the more formal Professorial Round. Her main areas of research were severe asthma, non-occupational and occupational interstitial lung diseases especially translational research, long before the term was in general use.

Dame Margaret managed to combine her medical and professional interests throughout her career and had a rich family life, having two daughters with her husband, Richard who had a 'remarkable'[vii] career as a urological surgeon. Both her daughters took good degrees, Lynne in music, she subsequently took up a career in medicine (Oxford and University College, Hospital, London), and is now Professor of rehabilitation at King's College London and Consultant Physician at Northwick Park Hospital. Gillian in science, and later ceramics. This chronological run through Dame Margaret's history makes the path look smooth and easy, but she would have been subject to the same strictures as most career women are. It is how she dealt with things that enabled her to keep everything on track. The work life balance is an important element in this.

The following extract from Dame Margaret's autobiography describes her experience of going up to Oxford, where at that time, only seven women were admitted to take medicine each year across the university out of 100 places.

This, she said, enabled her and fellow students to:

> "...gain experience, right from the beginning of working in an environment where women were in an extreme minority." and she gives good advice: "By far the best way to cope with this was to ignore gender attitudes altogether, even when snide remarks or evident prejudice was expressed. However well justified the anger and frustration at such attitudes might be, rising to the bait only exacerbates the situation. By contrast not responding to such aggravation allowed us to learn how to adapt in ways which could foster real equality. Not that we should take all the credit. There were a great many people around, both staff and students who were hugely supportive even when they had little experience of coeducation...[viii];

> Of present times she says: "it is rather sad to watch some women today rebuffing common male courtesy in order to demonstrate their equality. Why not appreciate a little male graciousness and enjoy it?
>
> On the other hand, the deliberate policy of systematically ignoring gender prejudice because it is irrelevant to medicine and other management situations has served well throughout the rest of my career, which, for most of the time, has continued to be substantially male dominated, the vast majority of whom were hugely supportive. This attitude has, I suspect, caused many 'glass ceilings' to simply disappear if they are ignored..."[ix]

But acceptance of some situations is necessary she counsels: "However some things in the early days were insurmountable and compromise had to be made…" It is clear that Dame Margaret knew when to compromise and found that this opened up opportunities as being known for flexibility so often does. Teamwork and interdisciplinary understanding are key aspects of success, as are "transferable skills",[x] using experiences as possible applications to others, for example as Chairman of the Board of Governors of Bedales School, dealing with the governing body members and the headmaster, Dame Margaret notes is a structure reflected under different titles in so many other areas, so, the Governing Body become trustees and the headmaster becomes Chief Executive Officer, as in any other organisation. Ways of seeing are important in career development.

Throughout her life, Dame Margaret balanced the different areas; her very happy marriage between two eminent and widely honoured people in different medical fields is a triumph. Dame Margaret and Richard, himself a CBE in 1991, worked hard both at their careers and family life. They carefully considered where they should live, their helpers, their children's schooling and how their leisure time should be spent. For example a large garden enabled them to be physically active but close to their work.[xi] Dame Margaret adds "any professional success she may have had was very largely due to the unfailing support from her husband and 2 daughters."[xii]

To further honour her long and distinguished career, a College centre was named after her, and the eponymous Margaret Turner-Warwick Chair of Thoracic medicine.

At the age of 90, Dame Margaret officially opened the Margaret Turner-Warwick Education Centre for the NHLI in the Guy Scadding Building at the Royal Brompton Campus, in April 2016.

> Dame Margaret writes in her autobiography that she: 'wished to show young women at the beginning of their careers and young men contemplating a long-term partnership with a woman doctor, that however arduous the work is, it is still possible to have a full family life with children as lifelong friends where neither parent has had to miss out on professional opportunities...'[xiii]

Endnotes

i The Marvel of the Lung and Human Responsibility A Great contempt of God's Good Gifts? The Harveian Oration of 1994 *Journal of the Royal College of Physicians London,* 28(6), 1994, 530–537.

ii The Annual Margaret Turner-Warwick Respiratory Lecture was founded in 2006 as a collaboration between the National Heart and Lung Institute and the Royal Brompton and Harefield NHS Foundation Trust. Retrieved from http://www1.imperial.ac.uk/nhli/respiratory/lecture/. Accessed 13 May 2016.

iii *Living Medicine Recollections and Reflections*, Margaret Turner-Warwick. Royal College of Physicians, 2005, p. 23.

iv "...We met at Oxford in 1943 as contemporary medical students and married in 1950 when we qualified — we have integrated and supported our full time professional lives ever since..." Richard Turner-Warwick CBE, Emeritus Surgeon UCL. The Canadian Journal of Urology. Permission *Can. J. Urol.* 2009, 16(6): 4883–4886. Retrieved from www.canjurol.com/html/free-articles/V16I06-02-Legends_Dec09.pdf. Accessed 14 July 2015.

v *op cit.* iv 2005, p. 58.

vi *Ibid.*, p. 143.

vii *Ibid.*, p. 315.

viii *Ibid.*, p. 22.

ix *Ibid.*, p. 23.

x *Ibid.*, p. 314.

xi *Ibid.*, p. 312.

xii Personal communication.

xiii *Ibid.*, p. 311, 303 words quoted.

Obstetrics and Gynaecology

Dame Alice Josephine Mary Taylor Barnes

Name and Post Nominals

Dame Alice Josephine Mary Taylor Barnes (1912–1999), DBA, FRCS, FRCP, FRCOG, DBE.

First woman Consultant Obstetrician, Charing Cross Hospital London 1954.

First woman President, British Medical Association 1939.

Specialism

Obstetrics and Gynaecology; Women's health; Family planning.

Career Awards

1939 FRCS; 1941 DM; 1952 FRCOG; 1967 FRCP; 1977 Hon FRCPI; 1979 Hon MD Liverpool; 1980 Hon DSc Leicester; 1981 Hon MD Southampton; 1990 Hon DSc Oxon; 1994 Hon FRCOG; Royal Society of Medicine Obstetrics and Gynaecology Dame Josephine Barnes Prize.

Awards

DBE 1974.

Professional Affiliations

First Woman President, British Medical Association 1939; Vice president, Royal College of Obstetricians and Gynaecologists 1974; President,[i] Association of Chartered Physiotherapists in Obstetrics and Gynaecology 1977–1995; President, Royal British Nurses' Association; President, Medical Women's Federation 1966–1967; President, Women's National Cancer Control Campaign 1974; President, Osler Club London 1988–1989[ii]; President, British Medical Association; Royal Medical Benevolent Fund.

Other Affiliations

Governor of Benenden School for 25 years; Fellow of King's College, London, Honorary Fellow Lady Margaret Hall, Oxford, Royal Holloway and Bedford New College; President of The Friends of the Girls' Public Day School Trust.

Lectures

Given by Her

Royal College of Obstetrics and Gynaecology 17th James Young Simpson Oration 1977.

Quote

> I was very confident. I've always been rather, perhaps, over-confident. I was a very confident person in those days. You know, it never occurred to me I'd fail in anything. There we are.[iii]

Born

18th August 1912 Sheringham Norfolk.

Died

28th December 1999.

Family Background

Father: Reverend Walter Wharton Barnes, Methodist Minister; Mother: Alice Mary née Ibbotson, Fellow of the Royal College of Organists (Second woman to be a Fellow); three brothers, one sister.

Marriage

Sir (Harold) (Seymour) Brian Warren (1914–1996) 1942 MD[iv] (knighted 1972) Divorced 1964 three children, Penny (1914–1996), Amanda, Anthony.

Education

Schooling

Oxford High School for Girls 1920s.

University

Undergraduate

Lady Margaret Hall Oxford.

Degree

BA (Hons in Physiology) 1934; BM BCh Oxford 1934–1937.

Postgraduate

Clinical training: University College Hospital Medical School London.

Degree

MA; DM 1941 Oxford.

Postgraduate Awards/scholarships

Goldschmid Scholarship.

Text Discussing Career and Other Comments

As Josephine Barnes' father was a Methodist Minister he would have been on the Circuit, which meant that as the Congregationalist system is, he moved around the country depending on where Ministers were required. This meant of course that the family went too, so Josephine's childhood was spent in five year chunks in Basingstoke, Exeter, Scarborough and Oxford. That the family lived in Oxford was due to her mother inheriting money which enabled them to move there and unusually, then for an adult, for her father to attend St Catherine's College and obtain a theological degree.

Josephine's family life was one of variety — her father being a minister and her mother a musician, Which meant that a broad range of people were involved with the family. It was from her mother that Josephine's love and understanding of music and in being a performer came. Her mother was "a very remarkable musician, who had studied at the Royal College of Music and was only the second woman Fellow of the Royal College of Organists.[v] She encouraged all her four children to take up music. Josephine was the eldest of the four children who all played an instrument and sang. Two of Josephine's brothers became lawyers and excelled in their careers as did she in hers, but always had time for music and relaxation, they all sang in the Bach Choir together.

The move to Oxford meant that Josephine attended the excellent and enlightened Oxford High for Girls School, where she discovered her love of science and was encouraged to pursue a scientific career. At the age of 13 Josephine told her mother that she was going to become a doctor. "It came from inside me" not prompted by anyone else, it was her vocation.[vi] It was a way to give service to the community, which was a lifelong need which she fulfilled through her outstanding career.

At this period of time, in the 1920s, it was much less usual for women to choose such a career and to successfully pursue the path and to the extent that she did within the profession. Josephine did come up against obstacles to her chosen career of medicine from the male bastions within it, but won through by her determination, intelligence and sheer professionalism.

At the time of her matriculation, there were annual quotas for the number of women who could take medicine and Josephine was only one of five women out of 53 medical students in 1930. Only she and one other medical student attended Lady Margaret Hall, which she says was a blessing as it meant her friends studied other subjects such as classics, English, French, History, Music, and so kept her horizons broad. She played Hockey for the University and gained a blue.

Her degree was mainly a science degree[vii] but she was obliged to take physiology, obtaining a first in her BA, she then had to continue by taking medical studies, before attending University College Hospital (UCH) Medical School to gain her clinical experience.

She describes several of her Oxford tutors as being rather detached from the students and their study needs — the anatomy tutor, Arthur Thomson had written a book only for art students for example, not medical students and he segregated women and men for anatomy dissections. Josephine was "saved" by her tutor Alice Carleton who supervised dissection of the cadaver she had been left alone with to dissect via a book in Thomson's studio.[viii] Alice Carleton was also a sound advisor and suggested Josephine take the Primary Fellowship of the Royal College of Surgeons (FRCS), an examination in anatomy and physiology, which formed the basis of surgical training. She gained this by the time she was 21 in 1923.

The 1920s and early 1930s saw an unusual cohort of staff in teaching hospitals, including UCH where Josephine took her clinical training. Nine staff were Fellows of the Royal Society and also "not to have an FRS if on the MRC was unthinkable".[ix] This of course changed with time as the profession opened out.

To gain her MD she wrote a thesis on pulmonary tuberculosis (TB) and pregnancy, because her tutors at UCH had an interest in this as she says: "…now that was interesting, you see, because it combined my medical training with my obstetric training. I could…, put the two together, so, you know, there was so much to it, that I enjoyed it".[x]

She spent a summer at a private sanatorium, Mundesley in Norfolk where TB patients were treated. Her work was comprehensive: she was the medical officer and radiologist and pathologist and pharmacist. "And what better training could anybody have had, it was wonderful."[xi] All this stood her in good stead for an overall understanding and practical experience of managing a medical workload.

Despite her superb marks and top qualifications, she found it so difficult as a woman to get a junior doctors job as House Officer at UCH that she obtained one at King Edward VII Hospital, Windsor, gaining the necessary surgical experience. She is not bitter about her experiences, but continued to develop her trajectory circumventing difficulties by always taking the sound advice given as to where to apply next.

Josephine describes the work at Windsor as amazing, as the hospital serviced the populations of Windsor and Slough she operated in general surgery and in obstetrics and undertook blood transfusions, in a very makeshift manner, and she and fellow students also carried out the Windsor coroners post-mortems.[xii]

She then returned to UCH with a job she prized, House Officer in the Obstetrics Unit, where she qualified MRCP in the late 1930s.

As to why she chose to specialise in Obstetrics she says:

> First of all, that you're dealing with pregnant women, you're dealing with normal women, doing a normal event, and I enjoyed that. And secondly, in gynaecology and part of the obstetrics, there's an element of surgery, and I always enjoyed operating.[xiii]

She was inspired by her tutor Francis Browne who developed antenatal care where previously there was none.

In her obstetrics and gynaecology role at UCH, she also had to "do the district" which meant visiting patients in their homes and as part of a "flying squad" which she set up, to attend births. Most births were at home at that period and the area encompassed some of the poorest homes, which was revelatory in social terms, and added to her stock of experience for use in her later career.

With WWII imminent and then breaking out in 1939, Josephine had thoughts of joining up as an army medic, but the BMA suggested her skills were more valuable to the home front taking on posts vacated by enlisted doctors, particularly in surgery. She was working for her final fellowship in surgery at that time at the London Hospital, so she continued to do this. Her former mentor at UCL suggested she spend six months as House Surgeon at the Samaritan Hospital for Women, then at the City of London Maternity Hospital, having a lucky escape from both as they were bombed just after she left. She then worked at Queen Charlotte's Maternity Hospital which then seemed to be run by three consultants and a registrar from a distance. She does report that astoundingly, the whole hospital was moved in one day from Marylebone to Hammersmith because of bombing.

She then worked at the Radcliffe Infirmary in Oxford and at Ruskin College, which had been turned into a maternity hospital for women from Southampton, a south coast town which was being heavily bombed. After a year she returned to London as she felt the Oxford atmosphere artificial and too remote in escaping the brunt of the war damage, whereas her brothers and sister were in the army, navy and admiralty, and her parents in Cardiff and she had run the gauntlet of air raids in London.

Josephine returned to UCH, the Elizabeth Garret Anderson Hospital, where she worked and continued her association for over 30 years, remaining there even after her marriage to Brian Warren in 1942, and the birth of the first of their three children in 1943. Nannies and cooks were employed by Josephine and Brian so that they could continue their medical work as Josephine felt it best to leave those who knew what they were doing to the childcare and concentrate on what she knew best as a physician and surgeon. But all were part of the family and Josephine always got up to give the night feeds for her young children, and when older took them to school.

Much of the war Josephine and Brian spent apart, as he enlisted in Royal Army Medical Corps but was seconded to the Grenadier Guards, rising to the rank of Major. He gave distinguished service, being mentioned in dispatches.

Postwar, Brian set up his own private practice in Belgravia London in the large house they were renovating as indeed did Josephine, and she retained this practice until well beyond retirement from her hospital post.

This was of course pre the National Health Service which was introduced in 1948. This meant that salaries were introduced, previously payment had been via the patients. When asked what difference this made to her Josephine replied:

> '…all my hospital work I was paid for, whereas before, I was what is called "Honorary". I wasn't paid for it. Even so 'one had to supplement it with private practice in order to educate children, and run the house and so on, but one was paid.' She says things just carried on as previously, although her husband refused to join the health service and continued in his private general practice.[xiv]

Josephine continued with her hospital work, indeed she was associated with the Elizabeth Garrett Anderson Hospital (UCH) for over 30 years; in an interview, when asked if she felt it more difficult at that time, 1954, for a woman she says:

> Yes, you see one of the things that I'm pleased about as far as my life goes, I was the first woman consultant on Charing Cross Hospital staff. This was quite a responsibility because I felt that I had to be very careful and if anything went wrong, if there was a disaster of any kind they'd say, 'oh well, this is what happens if you have women on the staff.

'So I did 23 years without a single operation death, which I think was quite good…'[xv]

As adjuncts to her hospital work, Josephine was involved in various committees and matters such as birth control in Britain, and in the third world, on which she had worked with WHO, and abortion. She was appalled by the horrific results of backstreet abortions, and saw that legalising it did away with the dangerous conditions and gave more control to the women concerned and brought them into a proper care system. She sat on the Lane Committee on the Working of the Abortion Act, 1971–1973. She also ran an abortion counselling service for several years.

She was a member of the Todd Committee (The Royal Commission on Medical Education) which reported in 1966 and 1968.

Between 1961–1987 she was a member of the Medical Defence Union, undertaking medico-legal work, which looked at cases in which doctors are being sued for negligence for example. Josephine says the title is a misnomer, it is not a union as in trade union, but a defence body.[xvi]

With her brothers being lawyers she was used to arguing cases, but insists that her side of things was medicine not the law.

As President of the Gynaecologists of Europe, she was involved in debates on midwives and their training and the bringing together of the separate areas of obstetrics and gynaecology in some countries. She represented the issue of women medics in the Medical Women's Federation as President, 1966–1967 and fought for screening in the Women's National Cancer Control Campaign of which she was President from 1974.

Sport was a recreational activity as a student, but again competitive — following her hockey playing for Oxford University and at county level, she played for UCH too. She was 50 years old before she gave up playing altogether.

Josephine has always had time — or made time — to relax through music and reading. On the advice of an Oxford Don she read one non-medical book a week and kept that pace up throughout her life.

As to music she played the piano and accompanied her children and grandchildren as they played their chosen instruments until they "got too good for her". She was a financial supporter of the Camberwell Pocket Opera Company.[xvii] She stopped choral singing herself at the age of 86!

Finally Josephine says:

> I think that, in a lifespan, there is a certain amount that you can achieve, and that's, I mean, I've achieved as much as I reckon I can. I wish I could achieve more, but obviously everybody does, but, I think that within the limits that you have in any life, I've achieved as much as anyone can.[xviii]

Endnotes

[i] Association of Chartered Physiotherapists in Women's Health renamed in 1994.

[ii] Osler Club — The Club was founded in the late 1920s by six young doctors led by Alfred White Franklin and Walter Bett. They had trained at Oxbridge and London and wanted to pursue the clinical, historical, moral and ethical principles that contributed to the performance of a "good doctor". They chose Sir William Osler, who had died a decade earlier, as the embodiment of their intentions and the Club was named after him with the permission of Lady Osler. http://www.osler.org.uk/.

[iii] British Library National Life Stories, LEADERS OF NATIONAL LIFE. Dame Josephine Barnes Interviewed by Rebecca Abrams C408/015. Dame Josephine Barnes, p. 21, C408/015 Tape 1 Side B (Track 0200A0).

[iv] Physician, Medical Officer Grenadier Guards WWII. Private general practitioner London. Personal Physician to Sir Edward Health Prime Minister 1970–1974.

[v] The Royal College of Physicians and Oxford Brookes University Medical Sciences Video Archives MSVA 044 Dame Josephine Barnes DBE in interview with Sir Gordon Wolstenholme Oxford, 6th June 1989 Video Oxford Brookes Barnes, Dame Josephine 06/06/89 044 GW Transcript p. 9.

[vi] *Ibid.*, p. 1.

[vii] *Ibid.*, iii p. 12, C408/015 Tape 1 Side A (Track 0100A0).

[viii] *Ibid.*, p. 12, C408/015 Tape 1 Side A (Track 0100A0).

[ix] *Op cit.*, iii p. 3.

[x] *Op cit.*, v Dame Josephine Barnes, p. 32, C408/015 Grenadier Tape 1 Side B (Track 0200A0).

[xi] *Ibid.*, p. 4.

[xii] *Ibid.*, p. 4.

[xiii] *Op cit* v Dame Josephine Barnes, p. 32, C408/015 Tape 1 Side B (Track 0200A0).

[xiv] *Ibid* Dame Josephine Barnes, p. 78, C408/015 Tape 3 Side B (Track 0600A0).

[xv] *Op cit.*, v. p. 10.

[xvi] *Ibid.*, Dame Josephine Barnes, p. 71, C408/015 Tape 3 Side A (Track 0500A0).

[xvii] Retrieved from: http://www.epoc.co/?q=node/9.

[xviii] *Ibid.*, Dame Josephine Barnes, p. 119, C408/015 Tape 4 Side B (Track 0800A0).

Lesley Regan

Name and Post Nominals

Lesley Regan (1957–)
Professor BM, MD, FRCOG.

First Woman in the UK to be Appointed Professor and Head of the Academic Department of Obstetrics and Gynaecology at Imperial College School of Medicine at St Mary's Paddington 1996.[i]

First Woman President of the Royal College of Obstetrics and Gynaecology (RCOG) in 64 years in 2016 and only the Second Woman President Royal College of Obstetrics and Gynaecology (RCOG) and the First who is also a mother (twin girls).

Specialism

Reproductive medicine, infertility, in recurrent miscarriage and the development of non-invasive fertility sparing therapies for uterine fibroids, menopause.

Career Roles

Registrar, Obstetrics and Gynaecology 1984 Addenbrookes Hospital Cambridge; Secondment Medical Research Council, Embryo and Gamete

Research Group on *in vitro* fertilisation; Teaching Fellow 1985, Director of Medical Studies at Girton Cambridge 1986–1990; Consultant and Senior Lecturer in Obstetrics and Gynaecology St Mary's Hospital; 1990 Professor, Obstetrics and Gynaecology; Professor and Head of Obstetrics and Gynaecology 1996– St Mary's Hospital now Imperial College Healthcare NHS Trust; Director of People and Culture within the Department of Surgery and Cancer Imperial College; Director, Recurrent Miscarriage Service St Mary's; Clinician Tommy's Miscarriage research Centre; President of the UK Association of Early Pregnancy Units; Advisor to the NICE Interventions Committee; Professional Member, Human Fertilisation and Embryology Authority and Research Licence Committee. Trustee of Wellbeing of Women (WOW), National Confidential Enquiry into Patient Outcome and Death (NCEPOD); Genesis Research Trust; Charity for Research into Miscarriage (CHARM); Advocacy Committee and the FIGO Women's Sexual and Reproductive Rights Committee. Vice President, Strategic Development 2014–2016; President, September 2016– RCOG.

Career Awards

2005 "Woman of Achievement Award" in recognition of her services to Reproductive Medicine; 2015 Julia Higgins Award for continued work leading the Athena SWAN academic opportunities committee Imperial College.

Quote

> It is a great honour to be elected President [of RCOG] and I am excited to be given the opportunity to lead my colleagues over the next three years… With the wealth of expertise available within RCOG's membership, I am confident we will build on our work to improve safety and quality in women's healthcare, provide excellence in training and education and be powerful advocates for women and their families, both in the UK and globally.'[ii]

Career Comments

Lesley's investigative and research skills, led her into her specialism. Whilst at the MRC she discovered similarities between infertility and recurrent miscarriage. She wrote her MD thesis on this area in 1989: *Epidemiology and Mechanisms of Sporadic and Recurrent Miscarriage*. As her career progressed, she developed interdisciplinary teams working across boundaries, seeing collaborative, or joined up working, as the best way to obtain advances in treatments. Her methods have resulted in a very high level of positive outcomes for the treatment of infertility and high risk pregnancies. Working with radiologists, she pioneered the use of magnetic resonance guided focussed ultrasound (MRgFUS) for uterine fibroids, avoiding the need for invasive surgery. Care of her patients is paramount to Lesley. She is a great communicator and has published two popular books on miscarriage pregnancy. She regularly writes and broadcasts on issues of women's health. Being a highly successful working mother of premature twins, and having four stepchildren, her empathy with her patients and her audiences can only be considered as the best.

Endnotes

[i] Imperial College reporter 27, 11–24 June 1996. Retrieved from http://www.imperial.ac.uk/publications/reporterarchive/0027/news01.htm.

[ii] Lesley Regan elected president of Royal College of Obstetrics and Gynaecology Imperial College News 24th May 2016 by Kathryn Johnson. Retrieved from http://www3.imperial.ac.uk/newsandeventspggrp/imperialcollege/medicine/departmentofsurgeryandcancer/newssummary/news_24-5-2016-14-24-47. Accessed 22 June 2016.

Sources: Lesley Regan Home page Imperial College website. Retrieved from http://www.imperial.ac.uk/people/l.regan. Accessed 22 June 2016.

Surgery

Averil Olive Mansfield

Name and Post Nominals

Averil Olive Mansfield (1937–)
CBE, MB, ChM.
First Woman Professor of surgery in the UK.
First Woman Chairman of a department of surgery in the UK.
First Elected Chairman, Court of Examiners Royal College of Surgeons of England.
Founding Chairwoman of Royal College of Surgeons Women in Surgical Training (WIST).

Specialism

Vascular Surgery and training; surgery to the carotid artery for the prevention of stroke; women in surgery.

Academic Qualifications and Awards

MB; ChM.

Career Roles

Consultant Surgeon: Liverpool 1972; Hillingdon 1980–1982; Hammersmith RPMS, at St Mary's Hospital Paddington 1982–2002; Honorary Surgeon Great Ormond Street Hospital for Sick Children; Lecturer in Surgery Liverpool University Medical School; Professor of Vascular Surgery St Mary's Hospital Medical School and Director of the Academic Surgical Unit (1993–2004) and Imperial College School of Medicine; Emeritus (Director, Academic Surgical Unit, 1993–1999); Associate Medical Director, St Mary's Hospital, 2002–2004.

Career Awards

Honorary MD University of Liverpool 1994, Honorary FRACS 1996; Honorary FACS 1998; CBE 1999; FDSRCS 2003; FRCP 2005; FRCPSG; Honorary DSc University of Lancaster 2010; Honorary FRSM 2010; Fellow Imperial College 2012; *6 Women Surgeons* by Jane Brettle 2009 Photographic exhibition by the Royal College of Surgeons; Voted "One of the 100 Women Who Have Changed the World" by *The Independent on Sunday* 2012[i]; *100 Leading Ladies* Portraits and Interviews with Influential Senior Women in Britain by Nancy Honey[ii] — In this interview Averil says her inspiration comes from the Lake District.

Awards

CBE 1998, for services to surgery and to women in medicine.

Endowed

Stroke Association's Jack and Averil (Mansfield) Bradley Fellowship Award 2016.

Professional Affiliations

Member of Council, 1990–2000, Founding Chairman, Women in Surgical Training (WiST),[iii] first Elected Chairman, Court of Examiners, 1990–1992,

Vice President, Royal College of Surgeons of England 1998–2000; President, Association of Surgeons of Great Britain and Ireland 1992–1993; President, Vascular Surgical Society of Great Britain and Ireland 1996–1997; President Section of Surgery, 1997–1998, Royal Society of Medicine; President, 2009–2010, Presidential address on the subject of Professionalism; Chairman, Board of Science 2010–2014, responsible for several publications importantly: *Drugs of Dependence*: *The Role of Medical Professionals* British Medical Association; Trustee, Vice Chairman Council 2001–2003; Chairman 2004–2009; Vice President, 2009– Stroke Association; Circulation Foundation Patron; Chairman, St Mary's Development Trust 2007–; Member, Board IDHDP 2015–.

Lectures Given by Her

Hunterian Lecture Royal College of Surgeons 1972; Bradshaw Lecture Royal College of Surgeons of England 2000[iv]; John Kinmonth Memorial Lecture Vascular Society Great Britain and Ireland 1989[v]; George Adlington Syme Oration Australasian College of Surgeons 1996.[vi]

Quote

> Find out if it's going to be for you … And then, just go for it and follow your dream!
>
> On graduating as a surgeon: I was absolutely in my element.[vii]

Born

21st June 1937 Blackpool, Lancashire.

Family Background

Father: Oxyacetylene Welder; Mother: a housewife.[viii]

Marriage

1987, John William Paulton Bradley (1927–2013) also a surgeon, three step children, grandchildren.

Education

Schooling

Collegiate School for Girls Blackpool.

University

Undergraduate

Liverpool University Medical School.

Degree

MB 1960.

Postgraduate

Liverpool University

Degrees

DSc; FRCS 1966; ChM 1972.

Text Discussing Career and Other Comments

Averil describes how she came to surgery as a career, suggesting an early interest in study and a great curiosity: "libraries are responsible for my career choice. How else would I have found out about anything without books?"[ix] Remember, Averil grew up in a time long before computers and the internet, at a time when libraries did hold the key to knowledge and information couldn't be accessed remotely. But she adapted as we all have to, as she said in 2003: "It [vascular surgery] was quite a new specialty when I qualified in 1960 in Liverpool. It's fascinating to see how it's changed; so much more is now possible and anaesthetic and diagnostic skills are almost beyond belief."[x]

> Looking back, Averil recalls expressing her intention to be, not only a doctor, but a surgeon, at about the age of eight, after reading a children's book on early surgical advances which described a doctor operating opening the chest unaware whether a patient could live after such a procedure. Averil was an only child and her mother laid down a local rule

> that she could only say that she wanted to be a surgeon to her two parents, she was permitted to say that she might become a doctor to wider members of the family, but that she must respond to inquiries from the rest of the world by saying that she wanted to be a nurse…
>
> Her mother had originally nurtured a desire to be a teacher but had been told that she was not strong enough, although disproving this perhaps by living to the age of eighty-three, and had met her father in Blackpool Leisure Gardens. He had left school at twelve and work was not always easy to find. Averil passed the eleven-plus and was educated at Blackpool Collegiate School for Girls where her musical ear was recognised in song and piano. Winning a local musical competition, her mother reminded her of the family rule and accordingly she appeared in the local paper with the header *Averil to nurse*. School and Headmistress were supportive and she passed the necessary three A Levels for medical school with a distinction in biology, her mother having written off for the A Level syllabus and ensured she covered the work. Medical school in Liverpool was a big step and one of her schoolteachers told her that if she were lonely in the hall of residence to go and find a piano and play it and she did just that and found herself surrounded by a friendly group.[xi]

The beginning of Averil's career was spent in Liverpool, where she held posts at the University and at the United Liverpool Hospitals, before moving south east and working in west London.

Averil became Consultant Surgeon at the Royal Liverpool Hospital in 1972, and was lecturer in surgery at the University of Liverpool, remaining in these posts until taking a Consultants job at Hillingdon Hospital for two years, 1980–1982, and then to the Royal Postgraduate Medical School (RPMS) and the Hammersmith Hospital, then St Mary's Hospital Paddington and so to Imperial College School of Medicine, when St Mary's merged with Imperial College in 1988.[xii] She was Postgraduate Sub-Dean St Mary's Hospital Medical School between 1987 and 1991 and Professor of Vascular Surgery in 1993.

Discussing becoming a professor during her time at St Mary's Averil says; "It was never my intention, and I never looked for an academic job. But when I got it, I had no idea how many things a professor has to do!"[xiii]

Another role Averil had was as honorary consultant in paediatric surgery and vascular surgery at Great Ormond Street Hospital for Sick Children.

Of her work she says: "I enjoyed outpatients very much — that unknown person with an unknown diagnosis was a wonderful challenge and opportunity. I taught students that they were terribly lucky, that the experience was like getting a Christmas parcel.[xiv]

Averil was innovative in her surgical and training work and led the field in vascular surgery for many years. She admits that surgery is seen as a man's world by some: "Surgery has a macho image." She mentions a comment by a man in a lift that women cannot be surgeons. "He simply didn't believe such a thing was possible. I had a laugh, said I understand how you feel, but the truth is, I am a surgeon. She has also said: "...you also need to let students discover the things they are suited to do. There's no reason why women aren't suited to surgery. You have to be decisive which demands that you have a certain amount of confidence in your makeup. You also need a fair amount of stamina." Averil comments that whatever your profession, "...to admit openly from time to time that you may need help and support is an adjustment in the mind set..",[xv] and is a view advisable in all careers.

She is mindful of the patients point of view: "Their confidence in us is at the centre of (the treatment) being a good experience not a bad one."[xvi]

One of the awards which made her "realise (she) had achieved something,"[xvii] was her election to the Presidency of the British Medical Association (BMA). She was President of two other medical organisations and an active member of others, holding various positions on them. Included in this is the Council of the Royal College of Surgeons, which is by peer election and her instigation of their WiST scheme.

Averil is one of those professors who will never fully retire; she may not operate now, she retired in 2002, but seems always available to promote medicine, surgery and good health. She has a very pragmatic approach to what she does and says: "I decided that when I talk, I would no longer talk about surgical operations, as you soon lose touch, and I would rather talk about things I know about."[xviii] She knows about a great deal, and remains involved with the Royal College of Surgeons, the BMA's Board of Science, for which she gave evidence to a Commons Select Committee in March 2012, Imperial College, where she is Chairman of the St Mary's Development Trust[xix] and Patron of the Imperial College School of Medicine Surgical Society[xx] and is a patron of the Circulation Foundation.[xxi]

As relaxation, Averil walks, including fundraising activities such as the coast-to-coast walk for the Royal College of Surgeons in 1999. She and her husband had also renovated a house in Cumbria over a 30 year period, though they lived in Buckinghamshire. Averil still plays the piano and she is learning to play the cello[xxii]: "very slowly indeed. I thought, playing one note at a time can't be difficult, but if you don't make the note properly, it sounds dreadful!" Averil got her medical notation in good order over her long career of dedication perseverance and a love of medicine — the plumbing of the vascular system: "It's all about piping. A blockage in the piping needs to be cleared etcetera. Just as she gives advice to her students, whom she took so much pleasure in teaching, it is clear that she takes her own advice: Follow the thing you enjoy, but you need to make an effort to see what's on offer. Delve a bit deeper. Find out if it's going to be for you. It will pay dividends later on. And then, just go for it and follow your dream!"[xxiii]

Endnotes

[i] *The Independent on Sunday* 7th March 2012.

[ii] *100 Leading Ladies Portraits and Interviews with Influential Senior Women in Britain*, Nancy Honey Photographer, Hattie Garlick Interviewer and writer Dewi Lewis Media Ltd.

[iii] Retrieved from: http://www.rcseng.ac.uk/media/medianews/Collegecelebrates-womensurgeons. Women in Surgical Training (WIST) was established by the Royal College of Surgeons in 1991 to help address the under-representation of women in the surgical profession.

[iv] Bradshaw Lectures are given at the Royal College of Physicians and the Royal College of Surgeons of England (here given biennially).

[v] 1989 Professor Averil O. Mansfield — *An artery and a Vein Dancing — The Management of Arteriovenous Malformation.*

[vi] Kinmonth Lecture Averil O. Mansfield *Surgery: A Suitable Career for Women.*

[vii] Averil Mansfield Interview by Aaron Bancil ICSM Gazette Spring Issue 2011.

[viii] Averil's early family life is discussed in this extract from a book about women Fellows of the Royal College of Surgeons of England: *The Fellowship of Women: Two Hundred Surgical Lives*, Margaret Ghilchik, 2011, p. 170.

[ix] *Op cit.*, vi.

[x] Imperial College *Reporter* 127, 26 February 2003.

[xi] *Op cit.*, viii.

[xii] Imperial College Medical School College Archives Publication.

[xiii] *Op cit.*, vi.

[xiv] *Op cit.*, vi.

[xv] Imperial College *Reporter* 127, 26 February 2003.

[xvi] *Ibid.*

[xvii] *Op cit.*, vii.

[xviii] *Op cit.*, vii.

[xix] Development Trust of St Mary's, The aims of the Trust are: To fund research and teaching relating to Imperial College and St Mary's Hospital.

[xx] Imperial College School of Medicine Surgical Society, part of ICSM Students Union.

[xxi] The Circulation Foundation the Vascular Charity: Funds and promotes research into the causes, treatment and prevention of vascular disease in the UK.

[xxii] *Op cit.*, vii.

[xxiii] *Op cit.*, vii.

Hospice Movement and Palliative Care

Dame Cicely Mary Strode Saunders

Name and Post Nominals

Dame Cicely Mary Strode Saunders (1918–2005), OM, DBE, BA, MBBS.

Pioneering Woman

Founder of the Hospice Movement, Physician, palliative care specialist.

Joined Imperial in

1951 as a Research Fellow, St Mary's Hospital Medical School.

Education

Roedean School; St Anne's College Oxford BA 1945; Diploma in public and Social Administration 1945; MBBS 1957 St Thomas's Hospital London.

Career Roles

Physician; Hospital Almoner; Research Fellow, St Mary's Hospital Medical School 1958; Founder of the Hospice Movement; Medical

Director, St Christopher's Hospice; Founder, Cicely Saunders International; The Cicely Saunders Institute.

Career Awards

Lambeth Degree 1977[i]; Dame British Empire 1980; BMA Gold Medal for Services to Medicine 1987; Onassis Prize for Services to Humanity 1989; Franklin D Roosevelt Four Freedoms Medal for Worship 2000; Templeton Prize for Progress in Religion; Order of Merit 1989.[ii]

About

Cicely Saunders overarching concern was for dignity in a pain free death and living up to death. She devoted herself and her career to this aim. She had a sustaining Christian faith that also had bearing on the development of her career and work. She was capable of very strong personal attachments to people who were suffering, but who bore this in an inspirational manner.

Although she wished to be a nurse she did as her parents required and attended St Anne's College Oxford in 1938 taking PPE. Her studies interrupted by WWII, in 1940, she left to become a nurse at St Thomas's Hospital London, retuning to take a war time degree in 1944, partly because of continued spinal problems. Her diploma in Public and Social Administration enabled her to train as a Hospital Almoner at St Thomas's hospital. The contacts she made as a volunteer at St Luke's Home for the Dying in Bayswater, decided her to train as a doctor in order to establish herself and her work with palliative care. She qualified as a doctor in 1957, taking a Research Fellowship at St Mary's and working at St Joseph's Hospice for the dying in Hackney. This enabled her to begin her influential work on a holistic view of pain control, and in which death was not seen as a medical failure. She began research into pain control and fundraising for her own hospice, which she created as a charity in 1961 and opened in 1967: St Christopher's Hospice Sydenham. Her broad vision became an international pattern for the developing hospice movement. Cicely retained the post of Director of St Christopher's until 1985, when she became Chairman of Council. She had placed

St Christopher's on an academic footing which encompassed research and training. In further developing this area, she founded Cicely Saunders International[iii] in 2002, part of which is the Cicely Saunders Institute of Palliative Care, Policy and Rehabilitation, based at King's College London.[iv]

Medical Ethics was also an area that Cicely was profoundly concerned with, working with the London Medical Group (later the Institute of Medical Ethics).[v]

Proselytizing for her work was something Cicely had always undertaken, lecturing and writing and publishing internationally. Between 1963 and 1969, she lectured influentially in America on her vision for end of life care, remaining in touch with the many friends and colleagues she made there.[vi] She always insisted that St Christopher's was not the only pattern, but that care needed to be provided to suit the geographical area, and the people.

Cicely married in 1980, Marian Bohusz-Szyszko (1901/2–1995) he was a Polish artist whom she had developed a friendship with after buying a religious painting from him for St Christopher's.

Quote

> You matter because you are you and you matter to the last moment of your life. We will do all we can to help you, not only to die peacefully but to live until you die.[vii]

Endnotes

[i] Lambeth Degrees are conferred by, and at, the Archbishop of Canterbury's discretion, for outstanding contributions to humanity. Cicely Saunders degree was the first in the field of medicine for 100 years.

[ii] Order of Merit, the sole gift of the Sovereign to 24 members at any one time, this rewards those who have achieved greatly in the arts, learning, literature and science. Retrieved from: https://www.royal.uk/queen-and-honours. Accessed 25 April 2016.

[iii] Cicely Saunders International to promote education in and development of palliative care. Retrieved from: http://cicelysaundersinternational.org/what-we-do/. Accessed 13 May 2016.

[iv] Retrieved from http://www.kcl.ac.uk/lsm/research/divisions/cicelysaunders/index.aspx. Accessed 20150925.

[v] Cicely Saunders Wikipedia. Retrieved from https://en.wikipedia.org/wiki/Cicely_Saunders. Accessed 20150925.

[vi] David Clark End of Life Studies Blog Glasgow University 21st July 2014. Retrieved from http://endoflifestudies.academicblogs.co.uk/cicely-saunders-the-1960s-and-the-usa/. Accessed 20150925.

[vii] *Nursing Times*, 72, 1976, pp. 1003–1005.

Medic and MP

Baroness Edith Clara Summerskill

Name and Post Nominals

Edith Clara Summerskill (1901–1980)
Baroness PC, CH, MRCS, LRCP.

Fifth woman Life Peer to be appointed.

Specialism

Medicine and Politics.

Joined Imperial in 1918–1924 as a student at Charing Cross Hospital Medical School (Preclinical training was undertaken at King's College London during that period).

Born

19th April 1901.

Died

4th February 1980.

Education

Eltham Hill Grammar School; King's College Medical School 1918 and Charing Cross Hospital Medical School MRCS and LRCP 1924.

Career Roles

Medical practitioner, private GP practice; MP (Labour party) West Fulham 1938–1955; Warrington 1955–1961; Councillor Middlesex County Green Lanes Constituency Tottenham, London 1934–1941; Member, Labour's national executive committee 1944–1958; Party Chairman 1954–1955. Parliamentary Secretary, Ministry of Food 1945. Minister for National Insurance 1950; Member, later President, Married Women's Association supporting her private member's bill for the Married Women's Property Act 1964.

Career Awards

Privy Councillor 1949; Companion of Honour 1966; Life Peerage 1961.

About

Edith Summerskill combined medicine and politics throughout her feisty career, along with marriage and motherhood. She very strongly believed in equality of opportunity. She saw that this should be in the home, in marriage, and in public life, education and job opportunities and also in provision for health care. She campaigned within and outside government as an MP. She was fighting for the nationalisation of health care from the 1930s, and was angry at Aneurin Bevan being credited with creating the National Health Service in 1945. Edith felt it had been developed through the work of herself and others, particularly in the Socialist Medical Association. During the Spanish Civil War, she was a member of "National Women's Appeal for Food for Spain", and visited the women and children's refugee camps. She advocated reform of the homosexuality laws in the 1960s, and legalisation of abortion. In her political career she travelled widely, to Communist China, the Soviet Union and also to

Australia and New Zealand. She opposed the Suez war and American entry to the Vietnam War, and nuclear armament.

Edith was born into a medical family, her father being a GP whom she would accompany on home visits. Witnessing the appalling living, health, working and educative conditions of many of his patients was the beginning of her socialist stance. Her father was a political radical although a Liberal and supported women's suffrage. Edith married a fellow medical school student, Edward Jeffrey Samuel (1895–1983) in 1925, setting up a medical practice together in North London 1928–1945. Their two children, Michael and Shirley took Edith's name of Summerskill, which she had retained on her marriage.

Sources: ODNB John Stewart, 'Summerskill, Edith Clara, Baroness Summerskill (1901–1980)', *Oxford Dictionary of National Biography*, Oxford University Press, 2004; online edn, Jan 2011. Retrieved from: http://www.oxforddnb.com.ezproxy.londonlibrary.co.uk/view/article/31734. Accessed 8 September 2015.

London School of Economics Archives. Retrieved from: http://archives.lse.ac.uk/Record.aspx?src=CalmView.Catalog&id=SUMMERSKILL.

www. Parliament.UK. Retrieved from: http://www.parliament.uk/about/living-heritage/evolutionofparliament/houseoflords/house-of-lords-reform/from-the-collections/from-t.

Medical Statistician and Sexual Health Research

Jane Wadsworth

Name and Post Nominals

Jane Wadsworth (1942–1997)
FRSS, MSc.

Specialism

Medical Statistician, Academic Sexual Health Study Pioneer.

Career Roles

Computer Programmer, 1976–1979; Scientific Officer; Department of Medical Physics, St Bartholomew's Hospital, London 1979–1983; Research Associate, Department of Child Health, Bristol University and Paediatric Research Unit, Exeter University; Lecturer in Medical Statistics 1983–1989, Senior Lecturer St Mary's Hospital Medical School 1989–1997.

Career Awards

The Jane Wadsworth Clinic for Sexual Health Jefferiss Wing at St Mary's Hospital named after her.

Career Affiliations

Fellow of the Royal Statistical Society 1976.

Education

MSc Medical Statistics London School of Hygiene and Tropical Medicine.

Married

1966 Michael Wadsworth Dissolved 1991, Children: one son one daughter.

About Her

Joining St Mary's in 1983, Jane took part in a number of clinical studies, including the National Childhood Encephalopathy and that on Pelvic Pain Syndrome. Her great skill was in interpreting data along with those who had gathered it, and in assisting their outcomes from it.

Jane undertook pioneering work in the epidemiology and statistics of sexual health, which led on from her work on patterns of the spread of HIV infection in the UK, the first such survey. Her 1994 study, Sexual Attitudes and Lifestyles was serialised in the Sunday Papers as *Sexual Behaviour in Britain*.

Physics

Meteorology

Elen Elaine Austin

Name and Post Nominals

Elen Elaine Austin (1895–1987)
FRGA, FRMetSoc.
The Elaine Austin Centenary Memorial Prize in Aeronautics 2007, Imperial College Award.

Specialism

Dynamical Meteorology, Climatology

Joined Imperial in 1921.

Education

BSc Mathematics and Natural Sciences 1918 Newnham College Cambridge Imperial College.

Career Roles

Demonstrator; Lecturer; Personal Assistant to Professor of Meteorology Sir William Napier Shaw; his co-author on *Manual of Meteorology*; Principal Scientific Officer, Meteorological Office.

Career Awards

FRGA, FRMetSoc.

About Her

Elaine was seconded by the Meteorological Office, part of the Air Ministry, to work as Napier Shaw's Personal Assistant from 1921 to 1924. She undertook Advanced study in Dynamical Meteorology whilst working for Shaw at Imperial College. She continued to work on their four volume Meteorological Handbook after returning to the Meteorological Office. There she also prepared handbooks on the weather over the oceans during WWII and worked on Climatology. Retiring as Principal Scientific Officer in 1955, she chose to take a less senior role and continued working for some years. She completed *A bibliography of the works of Sir Napier Shaw, FRS, 1854–1945* in 1955.

Daily Graphic, 18. 4. 25.

METEOROLOGICAL LECTURE ON A KENSINGTON ROOF.

Sir Napier Shaw, president of the International Meteorological Committee now meeting in London, telling the members all about the balloon which goes up with recording instruments from the roof of the Meteorological Office in Exhibition-road, South Kensington.

Reading from left to right:- Meseguer, Matteuzzi, Hesselberg, Wehrlé, Davies, Gamba, Simpson, Walker, Richardson, van Everdingen, Hergesell, Shaw, Gold, Molchanoff, Cave, EEA.

Meteorological Office South Kensington 1925. Elen Elaine Austin signed as EEA in her manuscript text, far right.

Physics

Daphne Frances Jackson

Namc and Post Nominals

Daphne Frances Jackson (1836–1991),
OBE, FRSA, FIEE, FInstP, FAPS.

First woman professor of physics in the UK.

Specialisms

Nuclear Physics; Medical Physics; Women in Science, Women returners in science.

Career Roles

Assistant Lecturer in Physics, 1960–1962; Lecturer Battersea College of Technology 1962–1966; Research Assistant Professor, University of Washington, 1963–1964; Reader in Nuclear Physics University of Surrey, 1966–1971; Professor, 1971, Head of Department of Physics 1971–1988, Dean Faculty of Mathematical and Physical science 1977–1980, Dean Faculty of Science, 1984–1988; Visiting Professorships: University of Maryland, 1970; University de Louvain, 1972; University of Lund, 1980–1982; Institute for Cancer Research of the Royal Marsden Hospital.

Career Awards

FRSA 1984; Liveryman Worshipful Company of Engineers, 1986–1991; Open University Doctorate 1987; University of Exeter Honorary DSc 1988; Loughborough University Honorary DSc 1990; FIEE; FInstP; FAPS; Institute of Physics: Blue Plaque[i] on residence.

Awards

OBE 1987.

Professional Affiliations

SERC Nuclear Physics Board, 1973–1976 Member Physics Committee Board 1974–1976, 1985–1988; DTI Metrology and Standards Requirements Board, 1973–1975; Industrial Education Advisory Committee, 1983–1988; NRPB, 1980–1985; DHA West Surrey and North East Hants, 1981–1991, Vice Chairman, 1983–1984; Regional Science Committee, 1976–1985; South West Thames Regional Research Committee, 1983–1987.

Coordinating Group for Development of Training for Women, MSC, 1983–1989; UGC Physics Sciences Sub-Committee, 1985–1989; BBC Science Consultative Group, 1987–1991; Department of Energy Advisory Council on Research and Development for Fuel and Power, 1988–1991; Meteorological Office Agency Board, 1990–1991; Schools Exams and Assessment Council, 1989–1991; Surrey University Press, Director 1984–1991 Chairman 1984–1987; Surrey Medical Imaging Systems Ltd, Director, 1986–1991; Women's Engineering Society Honorary Secretary 1988–1991, Vice President, 1981–1983; President 1983–1985; Vice President Institute of Physics, 1974–1978.

Lectures

In Her Name

IET Daphne Jackson Memorial Lecture.

Quotes

> If your colleagues know you well enough, they don't give any thought to your being male or female.[ii]

Born

3rd September 1936.

Died

8th February 1991.

Family

Father: Albert Henry Jackson engineering fitter and turner; Mother: Frances Ethel Elliott a dress maker and designer pre-marriage; Siblings: one brother.

Marriage

None.

Education

Schooling

Peterborough County Grammar School for Girls.

University

Undergraduate University

Imperial College.

Undergraduate Degree

BSc Physics, ARCS 1958.

Undergraduate Awards/scholarships

School Leavers 1955–1956; Award Peterborough Major Scholarship 1955–1958.

Postgraduate

Battersea College of Science and Technology.

Degree

Battersea College of Science and Technology PhD 1962. University of London DSc in 1970.

Text Discussing Career and Other Comments

Daphne was the second child and only daughter of Albert Henry and Frances Ethel Jackson. Both parents clearly had practical skills which Daphne seemed to follow, by pursuing a career in science. Her skills were evident in her early career in pure scientific research, where she excelled in the intellectual pursuit of nuclear physics. She then turned her hand and mind to the practical application of physics to medicine, again leading the field.

Her primary education was in Peterborough, at West Town Infants and West Town Junior Mixed, before going on to Peterborough County Grammar School for Girls. There she was an acknowledged all-rounder. Academically, being absorbed in her work and able to work independently, didn't detract from her sociability and she was also a Prefect 1951–1954 and Head Girl 1954. Throughout her school career she had demonstrated a great sense of responsibility, conscientiousness and reliability, besides originality. This along with her grasp of English and grammar made her ideal to contribute material to the school magazine and also help staff to edit it. Acting was another talent, and sport too; she was goalkeeper for the hockey first XI and played cricket for the school. She was commended for her valuable and enthusiastic service to the school, her maturity, sense of humour and a most interesting personality.[iii]

On leaving school, she attended Imperial College where, in 1955, she found herself one of only two women on the physics course of 99 students.[iv] This didn't deter her in the least, and she graduated BSc in 1958 with the Associateship of the Royal College of Science (ARCS).

During her time at Imperial her abilities became known to the nuclear physicist L.R.B. Elton[v] (father of Ben Elton the comedian), and he invited her to work with him at Battersea College of Science and

Technology[vi] as his first research student, and so her academic career began in earnest. She recorded her occupation on leaving Imperial College as: Research Assistant at £300 per annum, plus six hours teaching a week at 19 shillings or 24 shillings per hour. She took up an Assistant Lecturers post as she was writing her PhD, which she was awarded in 1962 and was also appointed Senior Lecturer. During 1963–1964, she also worked in America, as a visiting Research Assistant Professor with John Sanborn Blair,[vii] a distinguished Nuclear Physicist at Washington University.

Battersea College of Science and Technology became the University of Surrey in the expansion of higher education in the 1960s.[viii] So, in 1966 Daphne became the University of Surrey's first Reader in Nuclear Physics and also the leader of the nuclear physics group. Her biographers state: "given this lead [the group] grew into the largest and most strongly supported such group in the UK."[ix]

She later became interested in medical physics and her department became world famous for its two courses on the subject area.

In 1965, at the age of 29, Daphne became the youngest ever Fellow of the Institute of Physics, which was a measure of her stature within the discipline, as this was very young to be elected to this honour. She was Vice President 1974–1978.

Daphne also undertook many advisory roles in major UK organisations, governmental educational and non-governmental such as the Meteorological Agency, most were associated with science or physics.

Daphne retained a modesty about her ambitions and achievements throughout her career, yet she published around 80 articles and four books on nuclear physics, 55 articles on medical physics and its applications, and 46 relating to science and society.

For much of her life, Daphne devoted to herself to her scientific work in physics, but always with her other interests, such as encouraging women into science and engineering, which she saw as a recreation, but which actually she turned into a profession!

In a 1972 interview,[x] she suggests that a solution to getting women into science and giving them early training is not straight forward. Single sex schooling may hold women back as the standards of teaching and lab quality may not always be as good because of lack of funding. While coeducational schools may be able to provide better quality facilities and teaching, the loss could be lack of senior job opportunities for women teachers, and the denial to girls of "an important source of precept and inspiration"[xi] found in many girls schools. Daphne considered these difficulties throughout her life and career as a scientist. If society did not value gender equality then science and education wouldn't either, but she would do what she could to change this attitude by getting women into scientific education and work places.

She became President of the Women's Engineering Society[xii] in 1983, a post she retained until 1985. During her Presidency she took the opportunity to create the Women in Science and Engineering (WISE) organisation.[xiii]

Daphne's initial concern was with women returning to work after a break to have a family. This put them at a disadvantage in knowledge and new skills, particularly in science as the subject moves on so fast. She wanted women to be able to use their previous skills at re-entry into the work place and to build up their confidence and new skills over a two year period with a project. Her original plan of a Women Returner's Fellowship became a Trust in her name when she died. A fitting memorial to a woman so involved, and so eminent in science. The Daphne Jackson Trust[xiv] has now extended its remit to include men. Thus keeping up with the times, as society and familial circumstances have changed since Daphne began her work. Now each gender has a greater opportunity to work in the way that suits them and their lifestyle.

As Daphne's biographers say:

> Her genius was to identify a genuine need in society, to plan an effective way of meeting this need, and to carry out that plan, raising the substantial funds needed and proving her many sceptics wrong in the process. Official recognition of these and related efforts came when she was appointed OBE in 1987.[xv]

Daphne's life and work is testament to the Peterborough Grammar School Motto: *Non sibi sed deo et alteri* Not for ourselves, but for God and others.[xvi]

Endnotes

[i] Blue Plaque scheme commemorates significant people's homes after their death. Retrieved from http://openplaques.org/plaques/30588.

[ii] *First Woman Physics* Professor Daphne Jackson talks to Caroline Smith In Person Section *New Scientist* 21st September 1972, pp. 480–482.

[iii] School reference Archives, Imperial College B/Jackson.

[iv] In 1955 UG entry women= 2%, and in 2015 UG entry women=26.2%.

[v] Lewis Richard Benjamin Elton.

[vi] University of Surrey Archives Founded as Battersea Polytechnic Institute in 1891 designated a College of Advanced Technology in 1957 name was changed to Battersea College of Technology. In 1966 charter as University of Surrey moved to Guildford 1968. Retrieved from http://www.aim25.ac.uk/cgi-bin/vcdf/detail?coll_id=2229&inst_id=42. Accessed 1 July 2016.

[vii] John Sanborn Blair 1923–1982, Physics Department of the University of Washington (Seattle) in 1952, until his death in September 1982. Retrieved from http://digital.lib.washington.edu/findingaids/view?docId=BlairJohnSUA3314.xml;brand=defaul.

[viii] *Op cit.*, vi.

[ix] Jackson, Daphne Frances (1936–1991) Physicist and champion of women in science and engineering by R.C. Johnson and E.A., Johnson Oxford Dictionary National Biography.

[x] *Op cit.*, ii, p. 481.

[xi] *Op cit.*, ii, p. 481.

[xii] Women's Engineering Society. Retrieved from http://www.wes.org.uk/aboutwes. Accessed 1 July 2016.

[xiii] Women in Science and Engineering. Retrieved from https://www.wisecampaign.org.uk/about-us. Accessed 1 July 2016.

[xiv] The Daphne Jackson Trust. Retrieved from http://www.daphnejackson.org/. Accessed 1 July 2016.

[xv] *Op cit.*, ii.

[xvi] Retrieved from https://peterborocountyschool.wordpress.com/about/. Accessed 1 July 2016.

Lady Anne Thorne

Name and Post Nominals

LADY ANNE THORNE.

Lady Anne Thorne (1928–)
MA, DPhil, BA.

First Senior Tutor for Women.

Specialism

Atomic Physics, spectroscopy.

Career Roles

Assistant Lecturer; Lecturer; Senior Lecturer; Senior Research Fellow, Senior Research Investigator in Physics.

Professional Affiliations

European Physical Society, Harvard College Observatory, Institute of Physics, AWRE Harwell, the National Physical Laboratory.

Born

3rd October 1928 Lady Anne Pery.

Family Background

Father: The fifth Earl of Limerick, Edmund Colquhoun Pery, GBE CH KCB DSO; Mother: Angela Olivia Trotter GBE CHD; Siblings: two brothers.

Marriage

In 1959 to Lieutenant Colonel Peter Francis Thorne KCVO CBE ERD, then Deputy Sergeant at House of Commons, subsequently Sir Peter Thorne, Sergeant at Arms, died 2004. Four children.

Education

Early schooling

Attended six different schools, the last being North Foreland Lodge, Kent 1935–1945.

Pre University

Chelsea Polytechnic London, taking the London University Intermediate in Physics, Chemistry, Pure and Applied Mathematics 1945–1946.

University

Undergraduate

1947 to 1950 Oxford University, St Hugh's College Oxford 1947–1950.

Degree

BA Hons (Physics and Mathematics) 1950.

Post Graduate

Oxford University St Hugh's College 1950–1953.

Degree

Post Graduate: MA (Oxon) 1954; D.Phil 1954.

Awards

Moberly Senior Scholarship 1950–1951; Elizabeth Wordsworth 1951–1953.[i]

Text Discussing Career and Other Comments

At her various schools, Lady Anne says she was taught virtually no science, yet she decided she wanted to be a physicist.[ii] Showing determination and the sense of how to get what she wanted, which characterised her future career, she ensured that she got the right training for a university undergraduate place by attending Chelsea Polytechnic London.[iii] Here she took the University of London Intermediate exams in physics, chemistry and pure and applied maths, (now the equivalent of a first year foundation BSc). This enabled Lady Anne to attend St Hugh's College Oxford first as an undergraduate 1947–1950, and then a further four years as a Postgraduate.

In 1954, on graduating from Oxford University, Lady Anne took up a year's Research Fellowship, working with Professor Norman Ramsey at Harvard University, who won the Noble Prize for Physics in 1989.[iv]

Following this, she joined Imperial College as an Assistant Lecturer in the Physics Department between 1955 and 1956, when she was promoted to Lecturer in Physics, a post she held until 1968 and further promotion to Senior Lecturer, which she held until 1993. She retired in that year and became Senior Research Fellow in Physics, Imperial College and then Senior Research Investigator in Physics.

Lady Anne is a very positive person, a collegiate person, who saw it was necessary to put a lot in to get a lot out, however, she mentions a lack of women's networking opportunities — the sort of opportunities that were afforded to men through their clubs and informal meetings. This was particularly the case during the period of her early work and into the 1980s, when women did begin to set up networks.

The lack of support for women and family life was something Lady Anne worked to remedy. Her huge input to College life, and the many things she has undertaken in her career and personal life described here show how she did it.

Lady Anne was the first incumbent of the role of Senior Tutor for Women Students from 1958 to 1978, a position she took very seriously indeed. She reports that it was at an interesting time socially, many women particularly were first generation university students. They had excelled at school in a smaller group, but found themselves tried by being in larger groups of even brighter students in some cases, with very little experience of living away from home. She was working with these women students as the 1960s women's liberation and sexual freedom movements were developing. She mentions the Brook Advisory Centres[v] that were fortunately available to give advice to young women (and men), impartially and free of charge. She worked a great deal with students and was a Member of College Committees on Student Residence, being Deputy Chairman for two years and on the committees of the Medical Services, Student Welfare, Admissions Policy and Overseas Students, as Chairman of the latter for two years.

One great achievement in support networks is the Day Nursery[vi]; an innovation that Lady Anne engineered with other likeminded women: Sandra Dawson, Margaret Cunningham and Janice Yeadon. It took Lady Anne five years of persuasive work with the College authorities. It opened to the children of staff and students in 1970, and it flourishes today. She saw the need for it so that family life became part of work life. Not only women, but men would be enabled to work well without being concerned about child care. Although it was at that period still quite the norm for women to give up work once they had children, Lady Anne had four children born successively in 1961, 1963, 1965 and 1968, during this period her physics posts was officially three quarters time.

Setting up and overseeing the Day Nursery for its first 12 years as the Steering Committee Chairman was a collegiate thing to do, but was also a reflection of her personal experience as a scientist and a parent. Besides the College committee work mentioned, she undertook departmental committee work as a member of the Physics Department Teaching Committee and Staff–Student Committee for six years, and the first Chairman of the Non-Professorial Staff Group.

Lady Anne held scientific appointments with bodies relevant to her work in atomic spectroscopy simultaneously to her family life, lectureships, and, extramural work at Imperial. One such appointment was the European Physical Society,[vii] and she sat on the Board of its European Group for Atomic Spectroscopy between 1979 and 1985, being Chairman between 1982 and 1985. As a member of the Institute of Physics, between 1988 and 1992 sat on the editorial board of the Journal of Physics B (Atomic and Molecular Physics)[viii] and was a Committee member of the Spectroscopy Group for six years, and Chairman for two of those years. Since 1992 she has been an Associate of the Harvard College Observatory.[ix]

Throughout her career Lady Anne has undertaken Consultancies, notably with "AWRE Harwell,[x] the National Physical Laboratory[xi] and several different industrial firms , which have included various aspects of instrumentation, wavelength and radiometric standards, plasma diagnostics and atmospheric remote sensing."

Lady Anne lists her research work as being on spectroscopy of atoms and diatomicmolecules, using dispersive, interferometric, magnetic resonance and laser techniques, with applications to atomic and molecular structure, interatomic interaction and transition probabilities, laboratory spectroscopic measurements for atmospheric physics and astrophysics. Development of spectroscopic instrumentation for the above, in particular the extension of Fourier transform spectroscopy into the ultraviolet region.

Her publications include books: *Spectrophysics*, Chapman and Hall 1974, Second Edition 1988. *Spectrophysics, Principles and Applications*, by Anne Thorne, Ulf Litzen, Sveneric Johansson, (Springer Verlag, 1999), with contributions to other books on specialist spectroscopic topics. Various book reviews and about 85 refereed papers in scientific journals. Many conference presentations, both invited and contributed.

Lady Anne lists her main non-scientific voluntary work as being "concerned with education. I have served at various times on the Governing Boards of 8 different schools and Colleges in both the State and Independent sectors. I was Chairman of Governors of Ardingly College for 5 years."

Lady Anne held her title in her own right, being the daughter of an Earl, but marriage to Sir Peter Thorne meant she held the title Lady again, as Lady Thorne. The comment below is from an anonymous former student and appears on Yahoo Answers in a posting related to titles:

> This vaguely reminds me of a lecturer I had at college. Lady Anne Thorne was the daughter of an Irish Earl (which made her a Lady in her own right) and then she married Sir Peter Thorne, which made her a Lady again. She was a physicist and taught our 2nd year spectroscopy class, so with our mathematical minds, we liked the idea of calling her Lady squared![xii]

Endnotes

[i] Miss Elizabeth Wordsworth was founder of St Hughes College Oxford.
Miss Charlotte Anne Elizabeth Moberly was Principal 1886–1915.

[ii] Lady Anne Thorne Autobiographical Communication from which most information in the text is taken Archives Imperial College B/Thorne.

[iii] Retrieved from http://www.kingscollections.org/exhibitions/archives/studentdays/chelsea-college/history. Accessed 24 June 2016.

[iv] Norman Ramsay biographical information: Retrieved from http://www.nobelprize.org/nobel_prizes/physics/laureates/1989/ramsey-bio.html. Accessed 24 June 2016.

[v] Brook Advisory Centres. Retrieved from http://www.brook.org.uk/aboutbrook. Accessed 24 June 2016.

[vi] Now called The Early Years Education Centre takes children between six months and five years of age of Imperial College staff and students Monday to Friday. Retrieved from http://www.imperial.ac.uk/early-years/.
Other women staff involved: Dame Sandra Dawson (Sociology Unit Lecturer); Janice Yeadon (Librarian); Margaret Cunningham (Computing Department Lecturer).

[vii] Retrieved from http://www.eps.org/.

[viii] Retrieved from http://www.iop.org/; http://iopscience.iop.org/0953-4075.

[ix] Retrieved from http://www.cfa.harvard.edu/hco/

[x] Retrieved from http://en.wikipedia.org/wiki/Atomic_Energy_Research_Establishment.

[xi] Retrieved from http://www.npl.co.uk/

[xii] Retrieved from http://answers.yahoo.com/question/index?qid=20110303132923AANBwkn. Posted by 'The Dark Side' c.2010 *Not available.* Accessed 24 June 2016.

Physics and Electrical Engineering

Hertha Ayrton

Name and Post Nominals

Hertha Ayrton in her laboratory at her home

Hertha Ayrton (Phoebe Sarah Marks) (1854–1923) First woman to be nominated to become a Fellow of the Royal Society (not elected, ineligible as a married woman).

First woman to be a full member of Institution Electrical Engineers (IET) 1899.

First woman to deliver a paper at the Institution of Electrical Engineers (IET) 1899.

First woman to deliver a paper at The Royal Society 1904.

First woman awarded The Hughes Medal Royal Society 1906.

Specialism

Physics, Mathematics, Electrical Engineering.

Joined Imperial 1884.

Education

Privately, Girton College Cambridge; University of London; City and Guilds College.

Career Roles

Governess; Lecturer; Inventor (26 patents) e.g. sphygmomanometer, mathematical dividers; air movement, anti-gas fan, electrodes, arc lamps.

Career Awards

Institution of Electrical Engineers Premium 1899; Read a Paper 1904, Hughes Medal Royal Society 1906.

Affiliations

Institution Electrical Engineers (IET) 1899; British Association.

Born

28 April 1854 Portsea, England.

Died

26th August 1923 Sussex, England.

Family

Father: Levi Marks Polish Jew, fled persecution to England, Watchmaker; Mother: Alice Theresa née Moss, Seamstress. Related to the Hartog family, educationalists, whose children, cousins of Hertha, became scientists. Siblings: seven.

Marriage

1885 William Edward Ayrton (1847–1908) Physicist and Electrical Engineer; Children: one daughter Barbara Bodichon (1886–1950); one stepdaughter, Edith.[i]

Quote

> Professor William Ayrton, once said to her cousin, Dr Philip Hartog, you and I are able people, but Hertha is a genius.[ii]

About Her

Hertha Ayrton, known as Sarah Marks as a child, had an impecunious upbringing, but at the age of 8 was invited by Marion and Alphonse Hartog to attend the school they ran in north west London, and learn alongside their children. In this way she came to know her cousins, Numa and Marcus well, and they were advocates in her future scientific career. However, her first job, from the age of 16, was as a governess, in order to earn her keep and assist her mother financially. She also changed her name to Hertha after the goddess of the earth in Swinburne's eponymous poem, influenced in this by a friend. Nonetheless, a very apt choice. Her looks and demeanour were striking, as was her intellect. Throughout her life, Hertha made very good friends with several influential women as well as men. George Eliot, one such friend was influenced by her looks and voice in her creation of the character Mirah in her novel *Daniel Deronda*, in which a theme is her exploration of the Jewish faith. Eliot was also a friend of Barbara Leigh Smith Bodichon, one of the founders of Girton College who became one of Hertha's greatest patrons, financially assisting her education and her later research career.

Hertha took the Cambridge University Examination for Women in 1874, passing with honours in English and mathematics, and went up to Girton College in 1876. Examinations were stressful for her, but she was assisted in preparation for the Cambridge Tripos in 1881 by extra teaching from Richard T. Glazebrook.[iii]

As a student at Girton, Hertha started the Mathematical Society and the Girton College Fire Brigade (candles being the usual form of lighting then were a big fire risk) and whilst there invented a sphygmomanometer (a device for measuring blood pressure).

After graduation, she returned to London, and ever interested in extending her knowledge, attended evening classes in electrical engineering/physics at Finsbury Technical College — later a feeder institution for the City and Guilds College. Here she met as her tutor, William Edward Ayrton, Professor of Applied Physics (Electrical

Engineering) and subsequently, in 1885, they married. Their daughter Barbara was born in 1886.

Will Ayrton was always supportive of Hertha's work, both her teaching, (she taught physics to women), and her research. However, until she was left a legacy she was unable to pursue her own research due to domestic duties. Provision was made by her eminent friend, Barbara Bodichon, a housekeeper was employed and Hertha was able to develop her scientific research career.

1899 was a defining year for Hertha's career: she read a paper at the British Association meeting and demonstrated the electric arc at a Royal Society evening Conversazione. She presided over the science section of the International Congress of Women, where she spoke up for women's appropriateness for working in electrical industries.

Hertha and Will shared a lab at the top of their house, but he did not research the same topics, as he feared her work would be attributed to him. She made work on the electric arc[iv] her own, and investigated vortices in air and water and their effect on materials — e.g. sand ripples on the beach, turning to this work as she nursed Will at the seaside where they were staying for his health. Hertha also studied effective air disturbance, devising a fan against gasses in the trenches during WWI, and dissipation of fogs and noxious gases from mines.

Hertha joined the cause of suffragism, becoming a member of the Women's Social and Political Union in 1906. During marches, police apparently had orders not to arrest her or Elizabeth Garrett Anderson. They were both well known and supported in scientific circles.

Hertha felt she had to take her place with the marchers, and offered a safe house for suffragettes released from prison after hunger strikes. With other women graduates in academic dress, she joined the science section of the Coronation Suffrage Procession on 17th June 1911, preceding George V's Coronation Day by a week, so hoping to press their point.

Hertha's best advocacy campaign for women's rights was actually her scientific work, and in fighting for parity with men in the many scientific membership organisations that denied women membership for so long. Most, e.g. the Royal Society and the Chemical Society continued this ban until the 1920s.

Her scientific cousins supported women in science: Philip Hartog (Chemist and Educationalist) women's entry to the Chemical Society and Marcus Hartog (Professor of Natural History, Cork), entry for women members to The British Association general and sectional committees in 1890. Silvanus P. Thompson (Physicist, City and Guilds College) seconded this. His intervention, that at the recent International Electrical Congress in Paris, "undoubtedly the most remarkable paper was one read by a lady, an English female electrician", that is Hertha, reading a paper on the electric arc, helped them win that battle.

Hertha expanded her home laboratory after Will's death in 1901, and continued her researches which led to the award of the Hughes Medal by the Royal Society in 1906.

There was a great paradox in Hertha being eligible to be awarded the Hughes Medal, "for her experimental investigations on the electric arc, and also on sand ripples"[v] yet not being eligible for membership, that is a fellowship, as she was a married woman and therefore, "not a person in her own right," but until the Sex Disqualification (Removal) Act 1919, this was the case.

Hertha was very much a person in her own right, and this was recognised by many men in the scientific community, but unfortunately not by the rules and regulations, which many of them were so slow to change.

As a foot note, we can celebrate the fact that although only the second awarded to a woman, and 102 years apart, in 2008, the second Imperial woman was awarded the Hughes Medal:

Michele Dougherty,[vi] "for her innovative use of magnetic field data that led to discovery of an atmosphere around one of Saturn's moons and the way it revolutionised our view of the role of planetary moons in the Solar System" (see separate entry).

Endnotes

[i] Barbara Bodichon Ayrton-Gould (1886–1950) suffragist, Labour activist and MP, married the writer Gerald Gould (1885–1936). Their son was the artist Michael Ayrton (1921–1975).

One step daughter, from William's 1st marriage to his cousin Mathilda Chaplin (1846–1883) one of the first women physicians. Edith Ayrton Feminist author, married Israel Zangwill. (1864–1926) author proponent of Cultural Zionism. Their son was Oliver Zangwill FRS 1913–1987 professor of Psychology Cambridge University.

[ii] IET Archives Biographies: Hertha Ayrton Biographical information on Hertha Ayrton, first woman Member of the IEE, and her work on the electric arc and the Ayrton Flapper Fan. Retrieved from: http://www.theiet.org/resources/library/archives/biographies/ayrtonh.cfm. Accessed 15 July 2016.

[iii] First Professor of Aeronautical Engineering at Imperial College.

[iv] Hertha Ayrton, The Electric Arc (London: The Electrician Printing and Publishing Co., 1902).

[v] Hughes Medal Royal Society for original discoveries particularly related to electricity and magnetism or their applications.

[vi] See entry Michele Dougherty.

Space Physics

Michele Karen Dougherty

Name and Post Nominals

Michele Karen Dougherty
FRS.

Second woman at Imperial, and second woman, to be awarded the Hughes Medal 2008, 102 years after the first, Hertha Ayrton, in 1906.

Specialism

Space Physics and Atmospheric Physics; Space, Plasma and Climate Planetary magnetospheres, particularly those of Jupiter and Saturn; Solar wind interaction with planetary bodies. Leading unmanned exploratory missions to Saturn and Jupiter.

Career Roles

Lecturer, Professor (Royal Society Research Professorship 10 years 2014–2024), Principal Investigator for the magnetometer instruments for two major outer planetary space missions: the NASA Cassini spacecraft, which is in orbit around Saturn; and the European Space Agency JUICE spacecraft that is due to go into orbit around Ganymede, Jupiter's largest moon, in around 2030. Designed the Magnetometer instrumentation for the Cassini–Huygens spacecraft.

Career Awards

Hughes Medal Royal Society for innovative use of magnetic field data that led to discovery of an atmosphere around one of Saturn's moons and the way it revolutionised our view of the role of planetary moons in the Solar System 2008; Chree Medal and Prize by the Institute of Physics for her contributions to the field of planetary magnetic fields and atmospheres and their interactions with the solar wind, in particular through scientific leadership of the Cassini mission to Saturn and its moons 2007, UK's top 100 scientists Science Council 2014.

About Her

Michele was born in South Africa and was inspired to take up space science, because her father built a telescope in their garden, grinding the lens himself, Michele and her sister helped with mixing the concrete for the telescopes base. Although Maths was her subject, applied physics became more and more interesting to her and she was invited to help build instrumentation. This led to the creation of the instrumentation for the Cassini mission. Michele recounts two great moments in her career, both linked to the Cassini mission.

> The 1st was when we all went to the Jet Propulsion Lab to watch the spacecraft when it first went into orbit around Saturn, six and a half years after the spacecraft had launched.
>
> The 2nd was my team on Cassini was responsible for discovering an atmosphere on one of Saturn's moons. We saw some observations on two of the flybys past the moon and it looked like an atmosphere. We weren't sure, but we thought it was important that we try to go really close on the next fly-by. And so on the next flyby, instead of being a thousand kilometres away, we persuaded the Project team to take the spacecraft really close, at 170 km away from the moon.
>
> I watched the data coming back with my heart in my mouth because if we had messed up no one would have ever believed me again! So that was another of the exciting bits. And then we started putting all of the data together.
>
> The work of space science is a very long game — 15 years or so from initial development, making and installing the instruments and then

the long space flights taking many years to reach their destination. The constant tracking of the space craft, and the final moments of deployment of the instrumentation on arriving at the planet, and the anxiety about it all still being in working order, culminates in the excitement of the data being relayed back to earth.[i]

Endnote

[i] Interview with Michele Dougherty by Laura Gallagher Imperial College Communications October 2010. Retrieved from: http://www3.imperial.ac.uk/newsandeventspggrp/imperialcollege/newssummary/news_31-10-2014-12-34-18.

Physics and Climate Change

Joanna Dorothy Haigh

Name and Post Nominals

Joanna Dorothy Haigh (1954–)
CBE, FRS, FRMets, FCGI, MA, Msc, DPhil.

First woman Head of Physics Department Imperial College 2014.

Career Roles

Professor of Atmospheric Physics, 2009–2014, Head of Physics Department, 2014–; Co-Director, Grantham Research Institute for Climate Change; Director The Science and Solutions for a Changing Planet (SSCP) Doctoral Training Partnership Imperial College; President Royal Meteorological Society 2012–2014; UK Representative, International Association of Meteorology and Atmospheric Sciences (IAMAS).

Career Awards

Chree Medal and Prize Institute of Physics 2004. Adrian Gill Prize Royal Meteorological Society 2010. CBE for services to physics 2013.

Quote

> Advice for scientists at start of career: It is tough. You need to maintain an interest, work hard and not be afraid to ask for advice. Don't cry in a corner if things don't work.[i]

About Her

Joanna studies the influence of the Sun on the Earth's climate. Using data from satellites and modelling the processes, Joanna is helping to untangle the warming effects of greenhouse gases from those of natural variations in solar energy.

> She has transformed our ability to predict the behaviour of the atmosphere and climate by integrating ideas from physics to produce computationally fast yet accurate models. Methods developed by Joanna are now used by researchers worldwide and incorporate the finest details of climate processes and solar influence.[ii]

Endnotes

[i] Stephanie Walton Interview with Joanna Haigh Felix Imperial College students newspaper. Retrieved from: http://felixonline.co.uk/science/3579/bumper-year-for-physics-joanna-haigh/.

[ii] Royal Society Fellows. Retrieved from: https://royalsociety.org/people/joanna-haigh-11560/.

Nuclear Physics

Dame Susan Elizabeth Ion

Name and Post Nominals

Dame Susan Elizabeth Ion (1955–)
DBE, OBE, FRS, FREng, FIMMM, FINucE.

Pioneering woman in the nuclear industry 1979 — Sole non-US member of the US Department of Energy's Nuclear Energy Advisory Committee.

Specialism

Nuclear Energy.

Imperial College

As Student in Materials Science BSc, DIC, PhD Metallurgy in 1979; Fellow in 2005; Visiting Professor 2006–.

Career

British Nuclear Fuels Ltd (BNFL) Chemical Engineering Section 1979; Head Research and Development Fuel Division 1990–1992; Executive Director of Technology 1992–2006; Member Company Executive

1996–2006; Chairman of the UK Nuclear Innovation Research Advisory Board (NIRAB) 2014–.

Other Work

Government Council for Science and Technology, advisor and advocate for the nuclear energy industry 2004–2011; UK representative on the International Atomic Energy Agency Standing Advisory Group on Nuclear Energy; European Union Science and Technology Committee 1996–, Chair 2014–; Member PPARC 1994–2000; Member Council ESPRC 2005–2010 Chairman UK Fusion Advisory Board 2006–2012; Member Board University of Manchester 2004–; Deputy Chair 2015– Non-Executive Director Health and Safety Laboratory Health and Safety Executive 2005–2014; Honorary Professor University of Central Lancashire 2007–; Visiting Professor London Southbank University, 2011–; Nuclear Innovation and Research Advisory Board 2014–.

Awards

Hinton Medal for Outstanding Contribution to Nuclear Engineering INucE 1993; Order of the British Empire (OBE) 2002; One of 180 women invited to a "Women's Theme Day" lunch at Buckingham Palace in recognition of her contributions to the field of science and technology 2004; Dame Commander British Empire (DBE) 2010; BBC Woman's Hour Power List 2013; Association for Project Management Honorary Fellowship; Royal Academy of Engineering President's Medal 2014; Fellow of the Royal Society 2016.

Professional Associations

President, British Nuclear Society 2004–2006; Fellow, Royal Academy of Engineering 1996; Vice President 2002–2008.

Quote

> The nuclear industry has an important role in ensuring that the UK has a balanced and secure energy supply, while at the same time meeting the challenge of reducing our carbon dioxide emissions.[i]

Born

3rd February 1955 Susan Elizabeth Burrows.

School

Penwortham Girls Grammar School Preston Lancashire.

Married

1980 John Albert Ion; Lives in Lancashire.

Endnote

[i] Athena Lecture 2004: Nuclear power: A solution to climate change or a victim of its history? — Dr Sue Ion debates the future of nuclear energy. Retrieved from http://www.imperial.ac.uk/college.asp?P=5248. The Athena Lecture is an annual event held to celebrate the achievements of women in science.

It's a Family Affair
19th–20th Century

Trenham Reeks (1823-1879)
- Registrar and Curator of Museum of Practical Geology
- Royal School of Mines 1839-1879

  - Margaret Reeks (1855-1937)
    - Technical Artist/Draughtswoman, author
    - Royal School of Mines
  - Maria Ellen Reeks (1858-1929)
    - Occasionally assisted her sister Margaret with her work
  - Trenham Howard Reeks (1870-1938)
    - Studied Metallurgy Royal School of Mines

The Reeks Family

Father: Trenham Reeks (1823–1879).
Registrar and Curator of the Museum, Royal School of Mines.

Trenham Reeks (1823–1879)

Assistant Curator of the Museum of Economic (later Practical) Geology (MPG) 1839. Curator and Librarian and Registrar (RSM) 1851. Trenham's association with the geological world created by Sir Henry de la Beche, spanned 40 years. From the age of 16, Trenham had been involved with the development of these organisations, and also the British Geological Survey. Having studied chemistry and mineralogy at the Museum under Richard Phillips, he was eminently qualified for the positions he held during his distinguished career. Trenham was an expert on ceramics, and with Sir Henry, produced a catalogue of the MPG collections.

Married

Christiana C. Reeks c.1850s, three daughters: Mathilda, Margaret, Maria, one son: Trenham Howard.

Name

Margaret Reeks (1855–1937).

Specialism

Artist; Technical Drawing.

Career

Technical Artist and author.

Born

1855 London

Father: Trenham Reeks (1823–1879), Mother: Christiana Caroline née Howard, Siblings: Matilda (1853–1924), Maria Ellen, a sculptor (1858–1929); Trenham Howard Reeks (1860–1938).

Career and Life

Born into the London geological powerhouse developed by Sir Henry de la Beche, for which her father, Trenham Reeks worked all his life — the Museum of Practical Geology, the Geological Survey and then the Royal School of Mines (RSM), Margaret followed in his illustrative footsteps.

All Geological Survey staff were excellent illustrators, necessarily, as pre-photography, drawing was the only way to show the terrain and specimens. Margaret and one of her sisters, Maria Ellen Reeks, inherited some of their father's aptitude and were both artistic, the latter becoming a sculptor.

Margaret was illustrator and artist and possibly later, a sculptor, an art teacher at times, but mostly employed as an illustrator for the staff of the RSM. She completed large teaching charts for their classrooms and lectures, technical work sheets and diagrams, and made models and was working closely with her father. Her brother, Trenham Howard Reeks (1860—1938) studied metallurgy at the RSM from 1877 to 1880.

Margaret worked at the RSM from 1883 to 1928, and she also collaborated with the Natural History Museum and the Imperial Institute, on drawings of crystals, a complex matter, publishing her didactic book *Hints for Crystal Drawings* in 1908.[i] In this year, she asked for a change in her title from Draughtsman to Technical Artist, to better reflect her work and status. Her work was in great demand — sometimes she had so many requests that she was allowed assistance in the form of employment for her sister Maria, who worked under her direction.

In 1920, perhaps because all her life she had been steeped in the world of the RSM, Margaret was asked to write the history of the RSM to accompany the published Register of Associates and Old Students. This has proved to be a great reference work to the present day. Recording in detail the lives of staff members and the development of the RSM 1851–1914.[ii]

Margaret worked for College until into her 70s, the teaching staff considering her "fully up to the work".[iii]

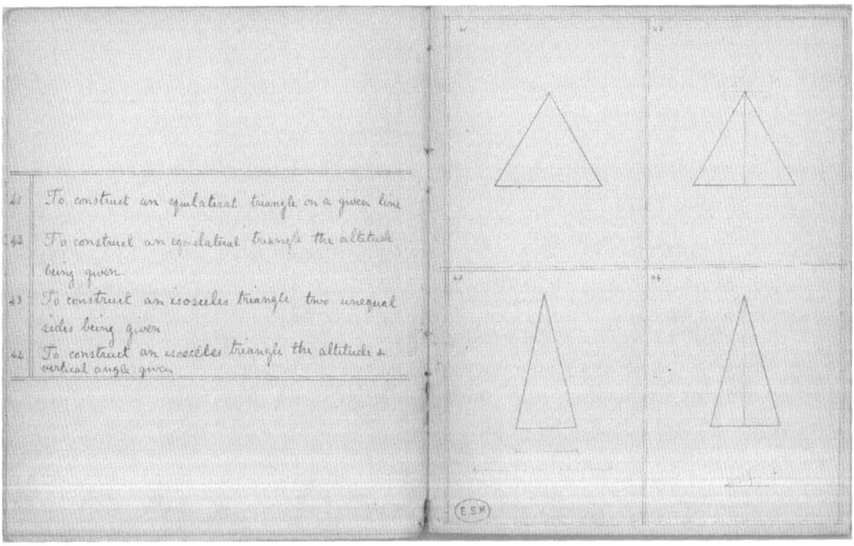

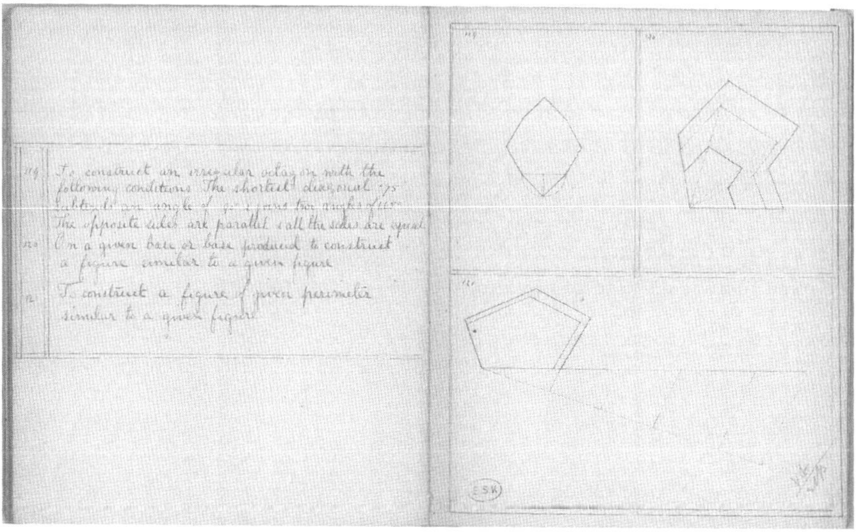

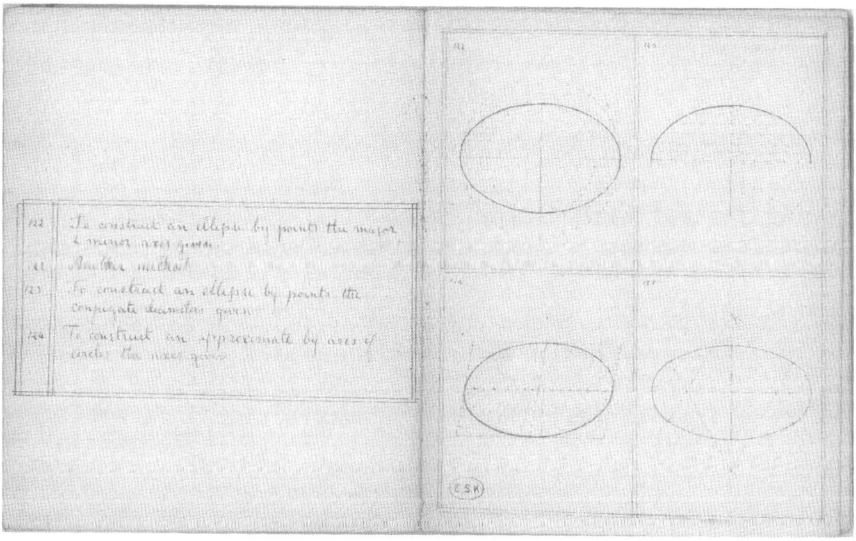

Images from Margaret Reeks' notebooks

Endnotes

[i] Hints for Crystal Drawing with a Preface by John W. Evans. with Illustration. by the Author, Reprint 2009. Retrieved from https://www.goodreads.com/book/show/8756704-hints-for-crystal-drawing-with-a-pref-by-john-w-evans-with-illus-by

[ii] Register of the Associates and Old Students of the Royal School of Mines and History of the Royal School of Mines by Miss Margaret Reeks 1920, London.

[iii] Departmental staff correspondence, Archives Imperial College.

The Judd Family

John Wesley Judd (1840-1916)
- Geological Survey 1867-1875
- Professor of Geology Imperial College (Royal School of Mines) 1876-1916

Hilda Mary Judd
- Studied Chemistry at Imperial College
- At Royal College of Science 1901-1904
- Organic Chemistry Research 1908 1914
- Biochemistry DIC 1915-1917
- Biochemistry Plant Physiology and Pathology Fruit Cold Storage Research 1918-

Father

John Wesley Judd (1840–1916)

Born 1840 Portsmouth; educated Camberwell; Royal School of Mines 1863–1864; Analytical Chemist Iron and Steel Works Sheffield 1864; Geological Survey England and Wales 1867–1870; Inspector of Schools 1871; Professor of Geology

1876–1905; FRS 1877; President of the Geological Society 1887–1888; CB 1895; Dean Royal College of Science 1895–1905; Emeritus Professor of Geology 1913, Imperial College.

Died

1916.

Name and Post Nominals

Hilda Mary Judd (1882–1951).
ARCS, BSc.

Specialism

Chemistry, Biochemistry research into silk and cold storage of fruits.

Career Roles

Research chemist; research into silk and cold storage of fruits Imperial College; Lecturer November 1906–August 1915, special subject Elementary Science Lecturer in Elementary Science 1907–1913. Lecturer in Nature Study and Botany 1913–1915 Goldsmiths College.

Awards

The Frank Hatton Prize 1904; Imperial Governors Free Place for Research Work in Organic Chemistry 1908–1909; 1909–1910, 1911–1912, 1915–1917.

Born

1882

Family Back Ground

Father: John Wesley Judd; Mother: Jeannie Frances née Jeyes; Siblings: brother.

Education

Private School Hastings 1897–1900; Royal College of Science 1901–1904; ARCS first Class (Chemistry) BSc (Science London) Occasional student 1908–1914; 1915–1917 DIC (Biochemistry).

Career and Life

Hilda undertook chemistry research work in chemistry at Imperial following graduation. Adventurously Hilda joined the British Association trip which met jointly in Africa with the South African Association in 1905. She left Imperial for a post at Goldsmith's College London between November 1906 and August 1915. She taught the special subject Elementary Science, becoming a Lecturer in Elementary Science 1907–1913, and Lecturer in Nature Study and Botany 1913–1915.[i]

Returning to Imperial, Hilda was appointed Research Assistant in the Department of Plant Physiology in 1916, working with Professors Blackman and Schryver on initial investigation into Silk. In 1918 she transferred to The Department of Plant Physiology and Pathology for work with Professor Farmer on Cold Storage Research. Here she also worked and published with Dorothy Haynes, both directed by Imperial's first woman Professor, Helen Kemp Porter.

Hilda became responsible for her ailing mother and brother after her father's death in 1916. As the need for her role as a carer increased, Hilda could not work. To alleviate the situation, the Institute of Chemistry arranged an allowance for her and Professor Judd's former pupils and colleagues set up a fund to provide her with an annual income — a family affair in need.

Died

1951

Tutors

Professor Vernon Herbert Blackman; Professor Samuel Barnett Schryver, Professor John Bretland Farmer; Martin Onslow Forster;

Martha Annie Whiteley; Dr Helen Kemp Porter, later first woman Professor at Imperial.

Endnote

[i] Goldsmith's Archives (Training Department Register).

Early 20th Century

Alexander Eugen Conrady
(1866-1944)
- Professor of Optics Imperial College (Royal College of Science) 1917-1931

Doris Conrady
(1904-1999)
- Studied Chemistry Imperial College
- RCS 1922-1925

Hilda Gertrude Conrady (1902-2003)
- Studied Optics Imperial College
- RCS 1920-1924

Irene Mary Conrady
(1905-2000)
- Studied Chemistry Imperial College
- RCS 1924-1925

The Conrady Family

Three of the four Conrady daughters attended Imperial College: Hilda, Irene, Doris. Hilda (Kingslake) is the one most is known about.

The Conrady Family in their garden Bedford Park West London. Rudolf Kingslake standing left

Father

Alexander Eugen Conrady (1866–1944).

Professor of Optical Design Physics Department Royal College of Science. Born in Germany. Studied science and maths at Bonn University Germany 1884–1886. Was diverted into mechanical industries and business 1890s discovered optical engineering, developed practical skills, designed instruments and lenses for an optical instrument maker W. Watson and Son, London.

Academic career: publishing papers on the microscope, telescope lenses and post WWI periscope lenses that he created. This led to his appointment as the first Professor to the newly formed Department of Optics Imperial College 1917–1931. Member, Royal Astronomical Society; Royal Microscopical Society; Quekett Microscopical Club, awarded Traill-Taylor Medal Royal Photographic Society 1920.

Died

16th June 1944, London.

Hilda Gertrude Conrady

BSc, ARCS, 1923, DIC 1924 (1902–2003)
later Mrs Rudolf Conrady Kingslake.

Specialism

Optical Engineering.

Career Roles

Research Assistant, Optics Imperial College; Optical Designing at Theodore Hamblin and Co; Researcher later Historian of Optical Institute University of Rochester New York. 1929–1990s.

Quote

> ...I inherited the optical genes, (a reference to her father's work, said playfully to her husband, also an optical scientist).

Born

16th February 1902, 104 Park Street, St Pancras London.

Died

February 2003, Rochester New York.

Education

1910–1920 Godolphin and Latymer School Hammersmith; 1920–1924 Imperial College.

Tutors

Professor F.C Cheshire; Professor A.E.Conrady; L.C. Martin.

Family Life

Married fellow Student Rudolf Kingslake 1929. They emigrated to America for work at The Institute of Optics at Rochester University New York. Rudolf as Professor of Optical Geometry and Optical Design, Hilda as a researcher. They retained close relationships with family in the UK, in visits and correspondence and offering accommodation and other help during WWII.

Children two sons, David Conrady Kingslake 1936 and Alan Horace Kingslake 1939. David died in a swimming accident in 1956.

Career

Hilda researched in Imperial's Optics Department after graduating, publishing regularly on the Foucault knife-edge test and primary spherical aberration. 1924 saw the publication of her first paper in the *Transactions of the Optical Society,* shortly after that of her fellow student and future husband Rudolf Kingslake.

1929 — Researcher then, later in life historian of, the Institute of Optics at Rochester University New York and of the discipline of optics. Hilda and Rudolf Published singly and jointly on optics, and wrote the second part of her father, W.A.E. Conrady's book, *Applied Optics and Optical Design* (1960) from the notes he had left. They jointly contributed hugely to the development of the Institute of Optics between 1929 and the 1990s. An important contribution to applied optics and its history is the article "*Alexander Eugen Conrady, 1866–1944*" that she wrote jointly with Rudolf, published in 1966 in the journal *Applied Optics*. Hilda's last publication was on the history of optics: a paper in Optics and Photonics News in 1991 entitled "*The First Institute of Optics in the World*." The work traced the early days of the formation and development of Imperial College's optics department, of which her father had such a pivotal role, and of which she and Rudolf had first-hand knowledge.

Conrady was the name Hilda retained in publications throughout her life, it being such a pedigree. However, Hilda and Rudolf were a partnership for over 70 years, dying within two weeks of each other in February 2003, Hilda aged 100 and Rudolf 99.

Doris Conrady

(1904–1999) (Sister of Hilda Conrady (Kingslake) and Irene Conrady). BSc, ARCS Chemistry 1925.

Specialism

Chemistry.

Career Roles

Junior Science Mistress Malvern Girls School Worcestershire 1925–

Born

12th March 1904 London

Died

1999.

Education

1912–1922 Godolphin and Latymer School Hammersmith; 1922–1925 Imperial College.

Tutors

Professor Herbert Brereton Baker and Assistant Professor Martha Annie Whiteley.

Undertook postgraduate research with Professor Baker.

Irene (Rene) Mary Conrady

(1905–2000) (Sister of Hilda Conrady (Kingslake) and Doris Conrady). BSC, ARCS 1928.

Specialism

Chemistry.

Career Roles

Freelance coaching in London Matriculation and University entrance exams in Chemistry, Physics Maths and Mechanics.

Born

24th August 1905 London.

Died

2001.

Education

Godolphin and Latymer School Hammersmith; Imperial College 1924–1928.

Life

Played hockey whilst at Imperial.

Mid-20th Century

The Hill Family Total of 116 Years' Service to the Geology Department

Edward (Ted) Hill (1910-1978)
Laboratory Boy/ Technician/
Departmental Superintendent
1925-1961

Violet (Peggy) Hill
Laboratory Woman/Technician/
Chief Technician
1943-1968

Allan Hill
Laboratory Boy/Technician/
Chief Technician
1946-1991

The Hill Family

Mr Edward (Ted) John Hill

Laboratory Boy paid 15 Shillings a week 1925.
Chief Technician 1946.
Departmental Superintendent 1961–1970.
Co-founder and Chairman Technicians Holland Club 1968.
Chaps Club member.
A keen apiarist and during WWII, placed two hives on RSM Building roof.

Mrs Violet (Peggy) Hill

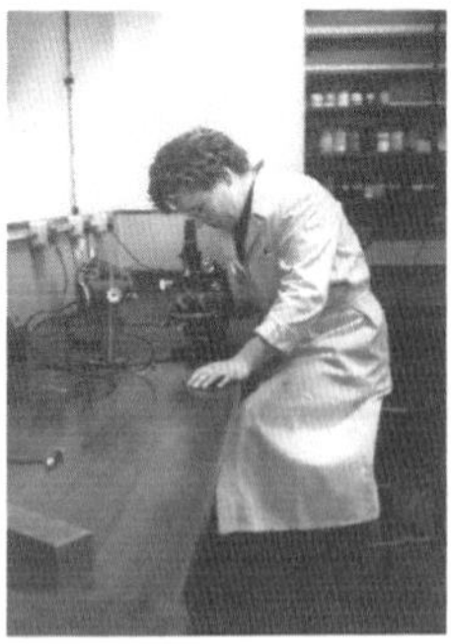

First Woman Technician Geology Department.

Quote

Professor Read (Head of Department) asked me one day how I was getting on, he said:
"You know you are the first woman to work in the Department in a technical capacity".
Temporary Laboratory Woman at £2 10s per week 1943.

Technician in charge of Ore Polishing 1966.
Chief Technician 1966–1968.

Allan Hill, son of Ted and Peggy

Laboratory Boy 1946–1947 at 25 shillings a week; General technician 1947–1959; Senior Technician Petrology 1959–1970 — curation of mineralogy and petrology collections and controller of this sections — 70,000 logged and labelled over the years. Chief technician 1970–1991 took early retirement.

Source: A History of the Geology Department Imperial College 1958–1988. Mary Pugh (Department Administrative Staff).

The Chain Family

Anne Ethel Beloff-Chain(1921-1991) - Lecturer in Biochemistry Imperial College London, 1964 - Professor of Biochemistry Imperial College London, 1983	Professor Sir Ernst Boris Chain (1906-1979) - Received Nobel Prize for work on penicillin, 1945 - Founded Biochemistry Department at Imperial College London, 1964

Imperial Male Contact: Husband

Professor Sir Ernst Boris Chain (1906–1979).

Anne Ethel Beloff-Chain
(1921–1991) (Lady Chain BSc, PhD).

Specialism

Carbohydrate Metabolism, Insulin role in diabetes and obesity.

Married

1948, Ernst Boris Chain Children: Two sons Benjamin and Daniel; one daughter Judith.

Anne and Ernst worked together at the International Centre for Chemical Microbiology at the *Istituto Superiore di Sanità* in Rome and in the Biochemistry Department instituted and headed by Ernst Chain at Imperial.

Anne Beloff Chain education and Career

BSc Chemistry University College London 1942; Biochemistry PhD Oxford University.

Postdoctoral work Harvard University

Chief Research Scientist International Centre for Chemical Microbiology at the *Istituto Superiore di Sanità* in Rome 1948–1964.

Lecturer, Biochemistry Department Imperial College 1964.

Professor of Biochemistry Imperial College 1983.

Professor and Head of Biochemistry Department University of Buckingham (Founded by her brother, Sir Max Beloff) 1985.

Professor Sir Ernst Boris Chain (1906–1979)

Nobel Prize Winner, FRS.

Born

19th June 1906.

Educated

Berlin Luisengymnasium, The Frederic-Wilhelm University.

Chemistry and Physiology 1930; DPhil Institute of Pathology Charité Hospital. University College Hospital Medical School Cambridge University, PhD 1933 .

Oxford, William Dunn School of Pathology to assist with the development of a Biochemistry Department, worked with the Professor of Pathology, Howard Florey 1935.

Nobel Prize for Physiology or Medicine jointly with Sir Howard Florey and Sir Alexander Fleming *'for the discovery of penicillin and its curative effect in various infectious diseases'* 1945.

Head of the Centre for Microbiology at Istituto Superiore di Sanitá Rome worked on semi-synthetic penicillins and fermentation technology 1950. Professor of Biochemistry and Head of Biochemistry Department Imperial College 1961, new department building with penthouse for the Chain family, Fermentation Pilot Plant. The Wolfson Foundation funding. Knighted in 1969; Emeritus Professor 1973, Senior Research Fellow; 1976 Retired moved to Mulranny Ireland.

Died

12th August 1979 Castlebar, Ireland.

20th–21st Century

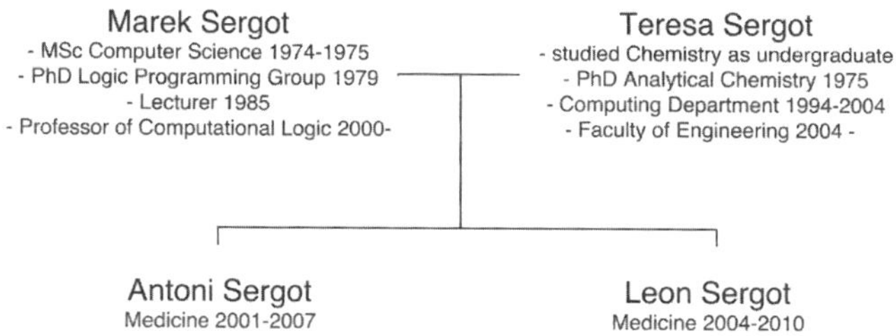

The Sergot Family

Marek Sergot

Marek seated front left. Teresa seated front right. Antoni standing left. Leon standing right.

Head of the Logic and Artificial Intelligence Section.
MSc in Computer Science 1974–1975.
Returned to Imperial began PhD in Logic Programming 1979.
Appointed Lecturer 1985.
Professor of Computational Logic 2000.
College Consul for the Faculty of Engineering and the Business School 2015.

Teresa Sergot

Undergraduate Chemistry Department 1968.
PhD in Analytical Chemistry 1975.
Computing Department in various roles: Academic Administration; Director Industrial Liaison overseeing Computing Department's Industrial Placement Programme for undergraduates 1994.
Chapter Manager Faculty of Engineering, supporting CGCA, RSMA and Engineering alumni 2004.
Academic and Alumni Relations Manager 2007–2016.

Two of their four children completed their medical degrees at Imperial College.

Antoni Sergot (2001–2007)

Leon Sergot (2004–2010)

The Hankin Family

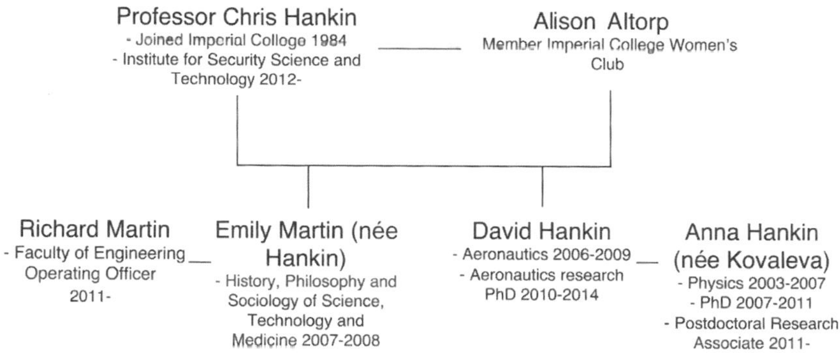

Chris Hankin

Chris Hankin joined Imperial College 1984.
Professor of Computing 1995.
Director, Institute for Security Science and Technology.
Dean City and Guilds College 2000–2003.
Pro-Rector for Research 2004–2006.
Deputy Principal of Engineering 2006–2008.
Research: theoretical computer science and language based security, multidisciplinary projects focussed on developing advanced visual analytics and decision support as defence against cyber-attacks.

UK Visual Analytics Consortium (UKVAC) cofunded by UK Government and US Department of Homeland Security.

Chair of the Lead Expert Group for the Government Office of Science's Policy Futures Foresight report on Future Identities.

President of the Scientific Council of INRIA, the French national institute for research in computer science and control, and Vice Chair of the DG CONNECT Advisory Forum for the European Commission.

Member, Cybersecurity Advisory Panel of the Global Cyber Security Capacity Centre.

2011 President CGCA Chris is very active in the City and Guilds College Association.

Alison Altorp

Married to Chris Hankin, mother of Emily and David, mother-in-law to Richard Martin and Anna Hankin.

Member, Imperial College Women's Club.

Emily Hankin

Married to Richard Martin.

MSc History of Science 2007–2008, currently on maternity leave.

Richard Martin

Married to Emily Hankin.

Management Trainee Imperial College 2000

Deputy Faculty of Engineering Operating Officer 2004

Faculty Manager, Faculty of Engineering Sciences at (UCL) University College London 2007 and subsequently had responsibility for establishing the university's first overseas campus in Adelaide as Chief Operating Officer of the UCL School of Energy and Resources, Australia

Faculty Operating Officer for the Faculty of Engineering 2011—responsible for leading the administration of the Faculty, delivering operational excellence and managing the Faculty's external relations initiatives.

Representing the Faculty on university committees formulating strategy, oversees planning and budgetary activities of the Faculty.

Project Director for the Aeronautics and Mechanical Engineering Project ("AMP") and the Dyson School of Design Engineering Project, a major refurbishment.

David Hankin

Married to Anna Hankin née Kovaleva.

Research Postgraduate Aeronautics Department 2010–2014 PhD. Investigation of the flow around a horizontal axis wind turbine operating in the wake of an upstream turbine in an unsteady aerodynamic environment, such as that found in large offshore arrays of wind turbines.

Research Engineer 2013–2014 Rutherford Appleton laboratory Aeroelastic modelling of a 5 MW horizontal axis wind turbine through coupling of an unsteady panel method with an Abaqus beam model.

Project Engineer 2014– BMT Fluid Mechanics Ltd.

Anna Hankin

MSci (Hons) Physics Department 2003–2007.

PhD Electrochemical recovery of nickel from nickel sulfamate effluents Chemical Engineering Department 2006–2011.

Postdoctoral Research Associate Imperial 2012–, Solar Hydrogen production; CO_2 reduction to Fuels; Pathways to Impact Solar Fuel Generators.

Left to right: David Hankin and Anna Hankin née Kovaleva, Chris Hankin and Alison Holtorp, Alison, wife of Chris and mother of Emily and David, member of Imperial College Women's Club; Emily Martin née Hankin and Richard Martin.

Liberal Studies and History of Science

Firsts: Women in Liberal Studies

Liberal studies as a subject area to back up the teaching of science and technology had been the concern of several Rectors before the establishment of the Touchstone series of lectures in 1950. This introduced eminent economists, and arts speakers to lecture at Imperial, and to take part in study weekends at Imperial's Field Station, Silwood Park near Ascot, Surrey.

One area of interest that had been an undercurrent for some individuals at Imperial for many years before the setting up of a department, was the History of Science. This was taught on a departmental subject basis by interested individuals such as G.J. Whitrow in Mathematics and Alec Skempton in Civil Engineering. They engaged women colleagues in this work during the 1960s, but to some extent as Whitrow also worked with his wife, Magda, they were in fact part of the 'Family Affair' scenarios mentioned elsewhere in this book. The leaders of the first History of Science Department were also part of the 'Family Affair' set. The department was set up in 1963, with Professor Rupert Hall (1920–2009) as Head of Department, and his wife, Maria Boas Hall (1919–2009), was Senior Lecturer. They were both historians of science in their own right.

Touchstone Weekend Silwood Park October 1959.[i]

Mathematics

Rosalind Cecilia (Cecily) Hildegard Tanner (1900–1992)[ii]

PhD 1929 Girton College Cambridge University.

Specialism

Mathematics, mathematical analysis and history of mathematics.

Mathematics Department Imperial 1933–1967, worked with G.J. Whitrow[iii] (1912–2000).

History of Mathematics

Their shared interest in the teaching and history of mathematics, led to setting up a seminar which ran from 1961–1967. At Imperial, she also worked with Sydney Chapman (1888–1970) FRS on a never completed text book. Known as Cecily, her main work was on the Renaissance mathematician, Thomas Harriot. She funded a Harriot seminar series in Oxford 1967–1983[iv] She was awarded the Gamble Prize Girton College for a mathematical history essay in 1939. Married in 1953, Bernard William Tanner (1881–1954) (Imperial College Chief Maintenance Engineer).

Magda Whitrow[v]

G.J. Whitrow's wife also had an interest in history of science and wrote a biography of Julius Wagner-Jauregg[vi] and translated articles. She spent many years based in the College library compiling the Isis Cumulative Bibliography. This remains a seminal work.

> Magda describes this as 'the high point of my career in the field of information science when, in the spring of 1984, 1 received from the publishers Mansell, the last volume, of the Isis Cumulative Bibliography (1923–65) which I had been editing. The Bibliography was based on references originally published in Isis, a journal for the history of science, which first appeared in 1913, edited by the distinguished historian of science, George Sarton. These 'Critical Bibliographies' were compiled from information or offprints supplied by colleagues from all over the world.'[vii]

Joyce Brown played a key role in the work on the history of the civil engineering profession. It was an interest of Professor Alec Skempton's, founder of the discipline of Soil Science.

Joyce had taught for some years after graduating from St Andrews University with a degree in English and history. Her first role at Imperial was in the Civil Engineering Department, as historical research assistant to 'Skem' as Alec, later Sir, Alec Skempton, was affectionately known. In 1972, ten years after her arrival, Joyce's role altered. In addition to her research role, she was appointed Lecturer to teach the history of the civil engineering profession and engineering economics, on the new

Environmental Impact of Engineering course. In 1979, Joyce became involved in another new course on Presentation of Information, comprising teaching of writing, drawing and public speaking. Joyce was in charge of the speaking and writing aspects of the course, and was also responsible for co-ordinating the teaching of communication skills to the undergraduates.

Joyce was the natural choice of the department to edit a book describing the development of Civil engineering at Imperial and its Constituent Colleges in the book: *A Hundred Years of Civil Engineering at South Kensington* in 1985. This reflected the founding of the City and Guilds College in South Kensington in 1885, which focuses on engineering.

Joyce's work with 'Skem', meant that she was immersed in the history of civil engineering, and the individuals concerned in the development of the profession, from its earliest times.

Pre, and post, retirement, Joyce edited *Studies in the History of Engineering* a 12 volume work on Civil Engineers.

To celebrate the Centenary of imperial College in 2007, a book about College from 1907–2007 was commissioned from a chemistry alumna, and Lecturer in History of Science at Simon Fraser University Canada, Hannah Gay.[viii]

Endnotes

[i] Imperial College Archives.

[ii] Born in Gottingen Parents William and Grace Young nėe Chisholm were academic mathematicians. Siblings: one brother Laurence Chisholm also a mathematician. Married 1953 William Tanner, formerly Chief Maintenance Engineer imperial, widowed 1954B. H. Neumann, 'Tanner , Rosalind Cecilia Hildegard (1900–1992)', *Oxford Dictionary of National Biography*, Oxford University Press, 2004 [http://www.oxforddnb.com.ezproxy2.londonlibrary.co.uk/view/article/54645, accessed 13 July 2016] Rosalind Cecilia Hildegard Tanner (1900–1992): doi:10.1093/ref:odnb/54645. Accessed 12 July 2016.

[iii] Gerald James Whitrow (1912–2000) Mathematician, Cosmologist, Historian and Philosopher of Science. One of his most famous books is *The Natural Philosophy of Time* (1960).

[iv] This ran until 1983. It is now annually held by Oriel College. Tanner financed a similar seminar at Durham University in 1977 which continues. Retrieved

from http://www.bbk.ac.uk/english/our-research/research_seminars/thomas-harriot-seminar/about-the-ths/the-ths-a-short-history. Accessed 26 June 2016, and Retrieved from http://www.dhst-whso.org/2013/01/neu-whitrow-bibliography-prize/, the prize commemorates the centenary of the *Isis Bibliography of History of Science*, which was started in 1913 by George Sarton. After Sarton, two bibliographers, John Neu at the University of Wisconsin, and Magda Whitrow at Imperial College, London, carried on Sarton's legacy. This prize recognizes the efforts of those two bibliographers for the work that they did to support history of science scholarship around the world.

v Magda Whitrow née Mostel, image from https://sites.google.com/site/cbibdoc/projects-and-prizes/neu-whithrow-prize/whitrow.

vi Julius Wagner, Jauregg Magda Whitrow (1857–1940). London: Smith-Gordon/ Niigata-Shi, Japan: Nishimura, 1993.

vii *Isis Cumulative Bibliography* A Bibliography of the History of Science formed from *Isis* Critical Bibliographies 1–90, 1913–1965 volumes 1–6. Retrieved from http://faculty.libsci.sc.edu/bob/ISP/whitrow2.htm.

viii *The History of Imperial College London 1907–2007 Higher Education and Research in Science, Technology and Medicine*, Imperial College Press 2007. Hannah Gay, Chemistry BSc 1961, PhD 1964 Imperial College; History of Science Sussex University; Lecturer Simon Fraser University; 2001–2013 Honorary Research Associate Centre for the History of Science Technology and Medicine; 2013– Honorary Research Associate Faculty of Life Sciences Department of Chemistry Imperial College.

Pastoral Roles

Imperial College has developed management of its responsibilities towards students and the minority that were women students since the early 20th century, and the beginning of the Women's Association. It has developed and formalised it over the years, as described here.

External Pastoral Assistance

In the 1970s, Imperial was working with external organisations such as the Women's Corona Society and the British Federation of University Women, particularly for the welfare of overseas students. The expertise of the women's Corona Society lay initially in assisting integration of diplomatic women into societies abroad and resettling them on their return to the UK. This was extended to other areas of life, including overseas students, and the wives of the male students. Their Hospitality Programme included home visits and a regular meeting of the English language circle, to assist with developing communication skills. The British Federation of University Women held International Tea Parties for postgraduate women at their Clubhouse, Crosby Hall in Chelsea.

Imperial College Pastoral Roles

There are many women who have played significant roles in the development of Imperial as a "good place to be" alongside their "day" jobs. They have made huge contributions to the pastoral and collegiate life of Imperial besides their academic work. This work continues, here some past and present contributors are celebrated.

In an informal audio interview in 2010, Margaret Cunningham, Anna Thomas-Betts, Marjorie De Reuck and Dot Griffiths described how they countered the gender imbalance, how they developed their own careers and that of pastoral roles such as College Tutor, and women's support groups.

Besides being early proponents of the work life balance, they did not see being a woman at Imperial should be a problem, but recognised that others perhaps saw women scientists and engineers as a special case. They wanted to instil a feeling of equality into College, without making too much of the gender issues. Life at Imperial was not to be about that, but about getting on with the job, as equals, in the lab and the field.

Others contributed to this ethos, such as Lynda White, (Mathematics) who held several such pastoral roles: Mathematics Admissions Tutor 1980–1987; Senior Tutor 1995–2011 1 of the four College Tutors from 2012.

Margaret Goodgame (Chemistry) as Undergraduate Admissions Tutor and a College Tutor 1993–1998, with particular responsibility for women,

(see her entry) and Lady Anne Thorne Physics (see her entry) Janice Yeadon (Central Library) and Sandra Dawson Management School (see her entry) supporters of the Day Nursery. Sue Thornett (opposite page centre of image) was the First Day Nursery Manager, she held the post for 27 years 1976–2003, overseeing its expansion and development into the Early Years Education Centre.

College Tutors

College Tutors advise on student welfare matters, reporting to the Dean of Students, Learning and Teaching.

Lynda White

Principal Teaching Fellow in Experimental Design Mathematics Department, she also designs experiments for other departments.

Roles

PhD student 1969, Lecturer, Senior Lecturer Statistics, Mathematics Admissions Tutor 1980–1987; Senior Tutor 1995–2011, one of the four College Tutors from 2012 PhD student 1969. President's Medal and Award for Excellence in Teaching 2008.

Margaret Cunningham

Honorary Senior Lecturer Computing Department retired in 2014 after 43 years at Imperial

Roles

Lecturer, Senior Lecturer, College Tutor.

In the citation for her Rector's Medal for Excellence in Pastoral Care in 2010, Margaret is recognised for the exceptional care she shows to students, offering sympathetic, non-judgemental counselling and willingness to make herself s available at very short notice to students who need advice. Noted is the thorough care that she provides for all students who need help, such as a student with deteriorating eyesight who said: "She did not let me give up and always provided practical solutions. I owe my future to so many people during my time at Imperial, but the key person who put everything together and made things happen was Margaret."[i]

Together Margaret and Lynda accepted a cheque for student prizes on the Joint Honours Mathematics and Computing Course in memory of a former Student Donald Watts Davies, who invented packet switching, vital for the functioning of the internet.

2004 Margaret Cunningham, (1st left) Computing Department and Dr Lynda White (2nd left), Mathematics Department receive a £15,000 cheque from Graham Turner and Andrew Glasspool of Farncombe Technology Ltd.[ii]

Anna Thomas-Betts

MBE, Senior Lecturer Geophysics College Tutor 1998–2010. Anna was part of the student support network so important to Imperial. She was surprised and delighted to be awarded an MBE, which she saw as a great honour and one she accepted as a token of acknowledgement and appreciation all those involved directly with student support. She also saw it as "a tribute to the women academics of the 1970s and 1980 who showed great staying power and I am sure many minority ethnic staff will take encouragement from it as well." Julia Buckingham Pro-Rector Education gave Anna the accolade of her contribution to the welfare of the students being "legendary."[iii]

Rector's Award for Excellence in Pastoral Care 2010.

Anna was brought up in India where her father was Physics Professor at the Union Christian College, Aluva and her Mother was Professor of English at St Xavier's College, Aluva. After undergraduate study in India, Anna began post graduate studies in the UK in 1962, joining Imperial in 1966 as Lecturer in Geophysics. She retired in 2009 after 43 years service to Imperial.

Anna Thomas-Betts 2009.
Photographer Karsten Seipp.

Margaret Goodgame Pastoral Work

Worked with women students in Chemistry and the College Tutors.
Undergraduate Admissions Tutor
College Tutor

Joined Imperial

1962 as Lecturer recruited by Geoffrey Wilkinson Inorganic Chemistry
Became Senior Lecturer.

Retired

1998

Education

BA, MA, D.Phil 1958 Oxford University.

Career

Assistant Lecturer Chemistry Northampton College of Advanced Technology London (Now City University) 1958–1959 and 1961–1962.

Research Associate MIT 1959–1961.

Lecturer in Chemistry Imperial College, with one year leave of absence teaching at University of Melbourne 1969–1970.

Awards

Associate Imperial College 2001.

Professional Associations

Member Royal Society of Chemistry.

Margaret Goodgame receiving her Associateship of Imperial College. Photographer Neville Miles.

The Rector's Awards for Excellence became the President's Awards for Excellence in 2014. These are annual awards to recognise staff members who have made outstanding contributions in Education, Research or Societal Engagement.

A great example of a recipient of the **Rector's Award for Excellence in Supporting the Student Experience** in 2011 is to be found in **Dr Lorraine Craig**, here opposite giving the Earth Science and Engineering welcome lecture to prospective undergraduates in an RSM Lecture Theatre in June 2014.

Lorraine raised the level of student satisfaction within the Department of Earth Science and Engineering to 98% in 2011, from 74% in 2008. Lorraine explains that the Department's active social life brings together staff in varied roles and students across all year groups. Bar nights, departmental lunches, sports games and pub quizzes help build connections between staff and students.

Lorraine says: "It's about smiling at students in the corridor or when they come to your office and not being afraid to show you care".[iv]

Lorraine is Academic Tutor and Associate Dean for Learning and Teaching, Faculty of Engineering, Department of Earth Science and Engineering. In addition to her role as Academic Tutor in Earth Science and Engineering, Lorraine works across the Faculty to support students

and staff in continuing to enhance the student experience from prearrival through to graduation. Lorraine deputises for the Faculty's Vice Dean for Education. She Chairs the Exploration Board which supports student expeditions and is a member of the Royal School of Mines Association. (RSMA) "My role is to be a liaison between the RSMA Committee and the current and former students, and staff of the Department of Earth Science and Engineering. As the Academic Tutor, I guess I get to know students (and some staff...) as well as anyone in the department. Although not an Imperial alumnus, earth science has been my way of life for a long time and mostly in the Arctic in recent years."[v] Lorraine is a geologist with a long term interest in the High Arctic, her first expedition there being made in 1984. She has led many expeditions and mapped the geology of areas of Greenland and Svalbard.

Imperial Coaching Academy

Another method of pastoral support is via a self-help system with a confidential coach, this is open to all staff, and many women in roles throughout Imperial have benefited from the programme.

Judy Barnett

Judy Barnett Talent Development Manager in the HR Division, explains: "The key thing about coaching is that it helps someone to think problems through for themselves — it's a completely nondirective approach. Coaches are trained to really listen and ask powerful questions in order to help their coachee carefully consider their options and come to their own conclusions and decisions." There are now around 40 coaches at Imperial, who are carefully matched with coachee before the pair agree on a timetable of four coaching sessions lasting around an hour each, spread over a number of weeks. All coaches attend a six day training programme which equips them with a coaching toolkit and a framework for appropriate, ethical behaviour. They are also offered continuing professional development (CPD) training twice a year.[vi]

The AWISE movement has been led by many of the aforementioned women and also others such as Sian Harding (NHLI) and Lesley Cohen Physics (see also entry first College Ambassadors).

Lesley Cohen

Professor, Solid State Physics Royal Society Research Fellowship 1995–2004, Imperial College Academic Diversity Taskforce committee member 2013 Silver SWAN award for the department of Physics 2009 (renewed 2012), IoP Juno Champion award for the department of Physics 2009 (renewed 2012).

Faculty of Natural Sciences Ambassador for Academic Women 2008.

Founder and Chair of Equality and Diversity Physics Juno Committee 2007.

Editor of booklet — "Voices from Women in Science" 1997.

Founder and Chair of Imperial Association for Women in (AWISE) STEM 1996.

The Association of Academic Women at Imperial (AWAIC) was founded by Sian Harding (NHLI) for women academics, postdocs and research staff from scientific and medical backgrounds. The inaugural meeting in focused on gender traits in grant applications behaviour.

Regular seminars on issues for women were held and AWAIC was a recognised part of the strategy of the Academic Opportunities Committee.

Sian Harding FAHA, FECISHR

Professor of Cardiac Pharmacology at the National Heart and Lung Institute (NHLI) Imperial College, heart failure researcher, cardiovascular biomedical research unit, Royal Brompton and Harefield NHS Trust, NHLI Lead for Women for Athena Swan, her interests include regenerative medicine and harnessing the immune system in this, and is Scientific Principal Investigator (PI) for the UK's first clinical trial on myocardial gene therapy; Co-PI in the UKRMP Immunomodulation Hub. Took part in the 2016 Royal Society's pairing scheme for MPs and scientists, with MP for Kensington Victoria Borwick spending a day in her lab.

Professor Sian Harding.
Photographer Jackie King.

PhD Pharmacology King's College London 1981. Joined the Cardiothoracic Institute in 1980 (later NHLI then Imperial College 1997) Professorial Inaugural Lecture, entitled *Reviving the Failing Heart: one cell at a time* 3rd April 2007.

Portrait taken as part of the Women@Imperial exhibition held on 13th March 2015.

Endnotes

[i] Retrieved from http://www3.imperial.ac.uk/newsandeventspggrp/imperialcollege/newssummary/news_2-6-2010-11-34-18. Accessed 9 July 2016.

[ii] IC *Reporter* 138, 10th March 2004.

[iii] Retrieved from http://www3.imperial.ac.uk/newsandeventspggrp/imperialcollege/newssummary/news_4-1-2008-13-1-36.

[iv] Imperial College *Reporter* 329, 3rd November 2011, pp. 8–9 Emily Ross-Joannou, Communica tions and Development.

[v] Retrieved from http://www.rsm-association.org/committee/committee.php.

[vi] Imperial College *Reporter* 293, 22nd March 2016, p. 10.

Imperial College Women's Clubs — Social

Wives Club

Set up post WWI it was run by wives of professors, Rectors and Chairman of the Governing Body.

It was fairly low key and involved mostly pastoral work for students and some visiting or overseas staff — teas and talks, but it thrived as an elite sort of group until the 1970s. Imperial was a very much smaller organisation until the expansion period of the 1960s, and also very formal. There seems to have been a form of hierarchy and professor's wives were part of that, Mrs Philip (Professor J. C. Philip Chemistry) Muriel Baker, a Chemist herself (Professor H.B. Baker Chemistry) along with participating wives of successive Rectors such as Lady Holland (Thomas Henry Holland Rector 1922–1929) and successive Governing Body Chairmen. For example, Margaret, Lady Falmouth, Hugh Boscawen Viscount Falmouth 1947–1962[i]) and Peggie Cooper,[ii] (Sir Frank Cooper 1988–1996) were active in the Wives Club from the 1950s.

The Women's Club

The successor to the Wives Club from about 1988, with a change in name, nature, outlook and management.

Professors' wives were notable for their involvement in keeping the clubs going, although some also had their own careers and most had children to bring up. These include: Helen Green (deceased) (Professor Mino Green Electrical Engineering), Rachel Caro (deceased) (Professor Colin Caro Bioengineering), Shirley Sargent (Professor Roger Sargent Chemical Engineering), Christine Stuart (Professor J. Trevor Stuart Mathematics), Glenys Pashley (Professor Don Pashley Materials), Helen Westcott (Professor John Westcott Electrical Engineering deceased), Helen Morton (Professor Igor Aleksander Electrical Engineering) Trisha Rees (Professor Charles W. Rees Chemistry deceased), Beryl Joyce (Professor Bruce Joyce Physics), Sheila Gibbon (Professor John Gibbon Maths Department), Pat Pendry (Professor Sir John Pendry Physics), Gail Bignell and Dr Ken Bignell (Physics.) Hilary Leppington (Professor Frank Leppington Mathematics Department and Principal of the Faculty of Physical Sciences 2002–2005). College Secretaries wives were also supporters: Sylvester Smith (John Smith College Secretary 1979–1989) and Babs Mitcheson a superb party caterer (Tony Mitcheson OBE College Secretary 1997–2007), many remain supporters of the Women's Club.

The role of President of the Women's Club was taken successively by Jean Vincent, (Lady Vincent,) wife of The Lord Vincent of Coleshill GBE, KCB, DSO, Chairman 1988–1996, and Elizabeth Kerr (Lady Kerr) wife of John, Lord Kerr of Kinlochard, Chairman 1997–2004. When Eliza Manningham-Buller became Chairman of the Governing Body, Professor Dame Julia Polak took on the role.

One of the Women's Clubs oldest roles has been support of the students' welfare, and an annual donation is made to a student support fund by the Club.

President of the Women of Imperial College club, Lady Vincent, presented a £350 cheque to the College Student Hardship Fund at a club meeting in December 1998. The cheque was accepted by Professor Julia Higgins, chair of the Fund, who commented, "The Student Hardship Fund is there to help IC students who face financial difficulty trying to complete their studies."[iii]

Imperial College Womens Club
Inspiring & Connecting Women of the College

The Women's Club is a networking club for all women staff, retired women staff and visiting staff and their partners. It is recognised by

College as part of the College staff support networks. The Club runs termly lunches with a speaker on a topic deemed to be of interest to women across College — male speakers are invited — and a summer soirėe with students from Imperial's neighbour, the Royal College of Music entertaining the guests. For members, the Women's Club holds a monthly Book Discussion Group (BDG) the book selection is wide and various, from 19th century French novelists to the latest prize winning novel, the BDG also engages with poetry and held a special meeting on the poetry of WWI to mark the 100th anniversary of the outbreak of war. Poetry is also read alongside prose works, and members are encouraged to share any poetry that, for them, resonates with the book of the month. There is also a Day Events Group which organises visits to places of interest and to exhibitions, for those able to attend during the day. The members Christmas lunch is always great fun and it is at this event that a collection is taken for the annual donation to the Student Opportunities Fund.

Lady Vincent presenting Dame Julia Stretton Higgins with a cheque from ICWC for the Student opportunities Fund.

Traditionally, the Rector's wife is the Chairman — consideration had been given to what would happen if there was no such partner — a senior member of the administrative or academic staff could be called upon to take the role, but having someone without other daily College responsibilities is preferable. This matter has arisen, as Imperial now has a woman President — Alice Gast, whose husband Brad Askins is supportive of the Club and its activities, but it is not a role for him! However, it is fortunate that now the College senior roles have split responsibilities, and as there is currently a male Provost, James Stirling, his wife, Paula Stirling, has agreed to take on the role of Chairman. Paula's earlier career was as a Lecturer in Law, and she retired from her role as Admissions Tutor at Cambridge University in 2014.

Rectors' Wives

Rectors' wives also had some input to the life of College in support of women, not only via the Wives or Women's Clubs, but on their own initiatives.

When Lord and Lady Flowers arrived — the then Sir, Brian as Rector (1973–1985), one of the duties that fell to Mary was, as she said, to run the Wives Club. She inherited "a very proper" organisation that was still using titles e.g. Mrs, Lady, instead of first names, but Mary did away with this standing on ceremony and loosened things up. Despite opposition about it being not quite, 'nice' she decided that fundraising would be one of the Club duties, to assist College causes such as the Day Nursery, saying:

"If we lose the day care centre we lose our clever girls."

Meaning that if women staff and students did not have facilities for child care, they could not work or study at Imperial.

To raise funds for replacement equipment or extras for the Day Nursery, Mary organised three fashion shows, enlisting help from the Royal College of Art, and produced a cookery book with recipes contributed by staff and the Rector. Sir Brian's recipe for punch is a cunningly low alcohol version, but pungent enough to suggest it wasn't, as he explained in their joint 2007 video interview!

In that interview, Sir Brian and Mary discuss using the Rector's residence to host "Beer and Bangers" parties — he would return home to find the kitchen full of Wives Club members, or students. He and Mary held twice termly "Beer and Bangers" parties in the Rector's residence for the students over the 12 years of their tenure, and the Student's Union threw one for them in return on the Queen's Lawn when they left College. Lord and Lady Flowers made the point that they were in the job together and they proved it by being so welcoming to their home and so comfortable with the students, who enjoyed their company. Mary was a particularly sparkling hostess. One of the student residences at the Silwood Park Field Station was named in her honour; "Mary Flowers" is a group of flats within the residences village.

Continuing the tradition of looking at doing the job together were Clare, Lady Ash and Sir Eric Ash (Rector 1985–1993). Clare set up the HUB, as an office dedicated to assisting the support networks for women and families of sabbaticals and academic visitors; organisations such as the Friends of Imperial, administered by Penny Berkut and the Women's Club were included. In the 1980s and 1990s it was far more likely that the academic visitor's dependents would be women, probably from overseas, with little or no idea about life in London, and few if any contacts. Clare wanted to make the difference in welfare terms to them, that the IC Union made to new students.

The HUB's support of family life and attempts to interest the young in science were made through events such as the Summer Fair and the Christmas Caper. The first Christmas Caper was an afternoon event in December 1988, it operated as a small-scale science fair intended to introduce children to science. St Mary's Medical School, newly joined with Imperial in 1988, provided entertainment for children through use of stethoscopes to hear their own heart beats. Other activities such as face painting and games were involved and staff were encouraged to bring their children. The Christmas Capers continued for some years, always including printing personal Christmas cards on Apple Macs and making Christmas wreaths. In some years, College combined with the two local Authorities, Westminster and the Royal Borough of Kensington and Chelsea Social Services, to invite disabled children to enjoy the afternoon. Gifts for the children were donated by staff and students.

In changing times in College, roles separated and although all subsequent Rectorial couples supported the College in many different ways, there was not to be a formal role for them. In the then Sir, Ronald Oxburgh's Rectorship (1993–2000), Ursula, Lady Oxburgh, herself a chemist and previously Head of Chemistry at the Perse School Cambridge, did take on the Chairmanship of the Women's Club, as did Janet, Lady Sykes who came to Imperial with her husband Sir Richard Sykes (Rector 2000–2008) Janet had a science degree from Bristol University in the 1960s. Rita, Lady O'Nions followed Janet as Chairman when her husband Sir Keith O'Nions became Rector 2010–2014. Rita's career had been in teaching. They like the Hills, Lady Hill, Helen, who was a violinist and Sir Roderick (Rector 1948–1954) many years before them, also supported the College music initiatives.

Endnotes

[i] Margaret Meynell (1894–1985) CBE, O.St.J. was the wife of Lord Falmouth (Evelyn Hugh Boscawen Viscount Falmouth Governing Body member 1932 to his Chairmanship between 1947 and 1962) they married in 1915. Lady Falmouth had associations with Swanley Horticultural College. During WWII she worked voluntarily to organize nursing and ambulance units for the front, for which she was appointed CBE and O.ST.J (Officer of the Most Venerable Order of the Hospital of St John of Jerusalem) in1946.

[ii] Peggie Cooper (née Claxton) d2004) wife of Sir Frank Cooper GCB, CMG, Chairman of the Governing Body 1988–1996.

[iii] Imperial college Reporter 20th January — 2nd February 1998. (There are currently a range of Faculty Student Union and College Support funds to assist Imperial's students.)

Imperial College Women's Association

Imperial College Women's Association (ICWA) Reminiscences of, Grace Payne née Lucas, President of ICWA 1944–1945[i]

This is a very personal view from alumna Grace, an RCS member at many years distance.

IC Women in the 1940s

Grace Payne

I was a student 1942–1945 and February–June 1946.

President of ICWA 1944–1945.

Number of women students was around 30, none in the RSM, around 6 in the C & G, the rest in RCS, variously studying Chem, Physics, Maths and Botany.

Working Hours were Mon–Fri 10 am–5 pm, no Saturdays. Lectures every day at 10 am followed by lab work till lunch time, some afternoons a lecture at 2 pm followed by labs, no spare time, though we were allowed time off for games/matches some afternoons.

In the Advanced Analytical Lab, 'Theo's Lab', [Leslie Stuart Theobald (1898–1979)] at the top of the building, men were arranged by surname in alphabetical order starting from A at the front, women were then placed behind the men so we got to know those at the end of the alphabet rather than those at the beginning. There were only a handful of us. On Friday afternoon all students were provided with a handful of cotton waste and some polish with which we were expected to clean our benches.

We sometimes slipped out during the afternoon for a tea break, somewhere across the road in the Imperial Institute Building[ii] but this was not approved by Theo!

Social Activities/Clubs/Sport. These include dances in the Union building, Dramatic Society, known as the 'Dramsoc', SCM (Student Christian Movement) — of which I was a member for some time, (London Interfaculty Christian Union) LIFCU. I don't remember any Debating Society, nor choir, nor orchestra (as there are now). Lots of sports — hockey, tennis, badminton, ice skating (at Queen's Club), I don't think there were any squash courts.

Hockey, tennis and badminton were played as mixed games, as there were insufficient women for women's teams. Hockey matches were played against teams from medical schools with the women players consisting of nurses. There weren't any medical schools which took both men and women students at that time, and all the other London colleges had been evacuated.

Tennis was played on the two courts in the quadrangle.

There was a Rifle Club but no women members — from choice, not that we were excluded.

Hostel. The only hostel, which was in the Beit Building, on Imperial Institute Rd. [Actually Students Union Prince Consort Road]. It was a small mixed hostel with one or two women residents, names which come to mind are Rosalie Hopkins and Joan Slow who were both in the C & G. Some of the men residents were Air Raid Wardens or Firewatchers.

Vacation Work. We were required to get what is now called 'work experience' during our summer vacations. I worked for one summer in the Pears' factory, in Isleworth, making experimental vanishing creams! The next summer in a small chemical manufacturing company somewhere near Hatton Garden.

Equality with men. We women were always treated as equals except that we weren't allowed in the Union Bar, (not until 1970) but I don't think that worried us. I can't remember any discrimination of any kind. Our only 'overseas' students were a number of 'refugees' from Nazi Germany/Austria who had mostly come to the UK shortly before the war started.

Dances were held in the Hall on the first floor of the main building, the back door gave access to the road at the back of the Albert Hall. At that time a popular dance was the Hokey-Cokey where a long single line was formed and one had ones hands around the waist of the next person and we would 'dance' right round the statue at the top of the steps, then back into the hall. It sounds 'corny' now, but it was fun. Women students from the Royal College of Music and some nurses from St George's Hospital, then at Hyde Park Corner, provided the necessary female input. Indeed, this provided at least one IC/RCM partnership resulted in marriage.

Travelling. Many students lived at home and travelled daily to college, others lived in digs locally. Some cycled, including me, and my bicycle was parked somewhere down a slope to the left of the Chemistry Department. There must at times have been problems with travelling when there had been bad bombing but it never seemed to be a big issue.

Promenade Concerts. A number of us were keen Prom-goers. We had a system whereby we bought a number of season tickets then shared them between us. At the end of the season we calculated how much each of us needed to contribute. It amounted to ten pence and three farthings one time, I remember, that would be approx 4p in present day terms! The 1944 season was aborted because of flying bombs. I don't remember how we sorted that out.

Constituent College Unions. There was both a Royal College of Science (RCS) Union and Imperial College Union, run by students who were elected by the student body. Both were non-political, concerned mainly with student welfare and social activities. Douglas (Payne Grace's husband) was Secretary of the RCS Union 1944–1945, followed by chairman 1945–1946, at the same time he was secretary of the IC Union. We were both much involved with student affairs.

Women's Common Room. I have no memory of a room set aside for women in the RCS building[iii] but we had our own room in the corner of the quadrangle on the Main building, where at some time we were able to buy new curtains, can't remember where we got them nor how we managed it as rationing meant that new fabric was difficult to come by. They were a wishy-washy fawn colour!

Endnotes

[i] I Grace Payne sent her reminiscences to Imperial College Archives in 2014 when she requested a copy of the article, *ICWA Then and Now* by Marion Gossett who was President in 1924.

[ii] Demolished 1957–1965.

[iii] Royal College of Science Building, except Chemistry 1 demolished 1974.

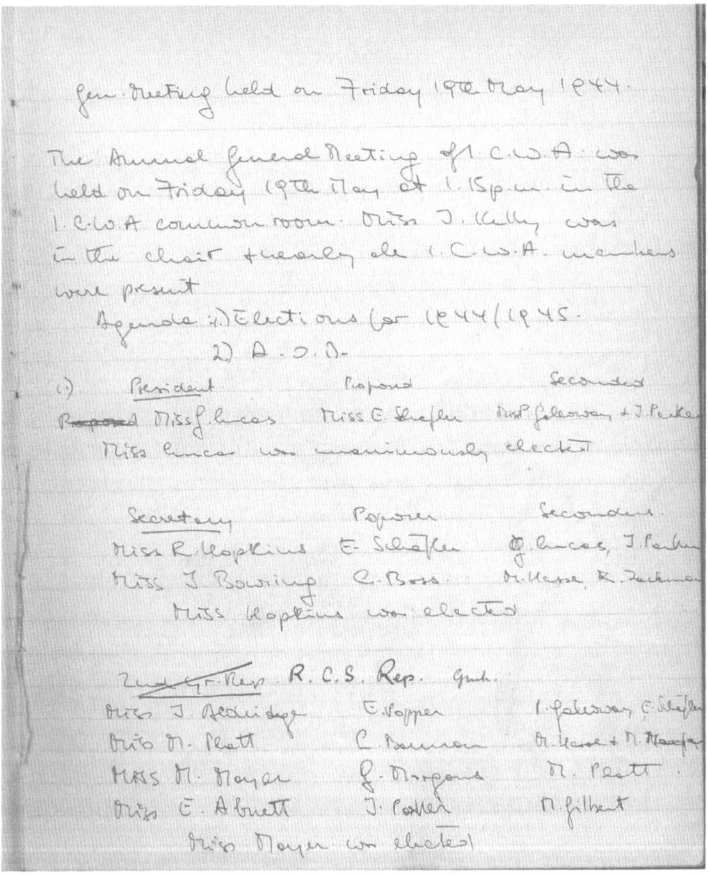

Gen. Meeting held on Friday 19th May 1944.

The Annual General Meeting of I.C.W.A. was held on Friday 19th May at 1.15 p.m. in the I.C.W.A. common room. Miss J. Kelly was in the chair & nearly all I.C.W.A. members were present.

Agenda 1) Elections for 1944/1945.
2) A.O.B.

1) President	Proposed	Seconded
~~Proposed~~ Miss G. Lucas	Miss E. Schafer	Miss P. Galloway & J. Parker

Miss Lucas was unanimously elected

Secretary	Proposer	Seconder
Miss R. Hopkins	E. Schafer	G. Lucas, J. Parker
Miss J. Bowring	C. Boss.	M. Hesse, R. Jackman

Miss Hopkins was elected

~~2nd Yr. Rep.~~ R.C.S. Rep. Grad.		
Miss J. Aldridge	E. Popper	I. Galloway, C. Schafer
Miss M. Platt	C. Bauman	M. Hesse & M. Moyer
Miss M. Moyer	G. Morgan	M. Platt
Miss E. Abnett	J. Potter	M. Gilbert

Miss Moyer was elected

Excerpt from ICWA General Meeting 19th May 1944, Miss G. Lucas is named as President.

I.C.W.A. Then and Now

During his speech at the Annual Dinner last year the President said that he thought ICWA might have been founded ten years ago; at this point Miss Marion Gosset stood up and interjected "I was President fifty years ago!" Miss Gosset pursued a career in Librarianship; her first job after a year's research was to the Science Museum Library where at first she was appointed to compile a list of periodicals. She expected to stay only a few months but actually stayed twenty years, becoming head of the cataloging department. She left in 1946 to create the library at A.E.R.E. Harwell, which has built up the outstanding collection of the world's literature on Atomic Engery with a staff of over fifty.

We asked Miss Gosset to write about the ICWA she knew and so, with the help of some contemporaries, she has written:–

I.C.W.A. in 1924

In our day the Association was not known as "Icwa", acronyms not having taken hold of life, but I think as the Women's Ass! It was very informal, so much so that in the Minutes of the I.C. Union Committee of the 29th October 1923, we read *"A letter from the I.C. Women's Association was then read .... It stated that as no rules had previously been submitted by the Association ... they now wished to submit the enclosed set and asked that this Committee should adopt them".* This they did. The rules were very simple covering Name and Object, Membership, Officers and Committee. One member of the Committee to be the Representative on the I.C. Union Committee. The last two paragraphs are interesting:

"The Committee shall undertake charge and control of any rooms allotted to women students in the several buildings of the College".

"If there are six or more women students in any one of the Colleges they are entitled to send one representative to the students' Committee of the College".

We had regular Committee meetings and an Annual Meeting and we kept Minutes, but alas all trace of the Minutes has been lost.

There were about fifty members in 1924. The Queen Bee was undoubtedly Dr. (afterwards Professor) M.A. Whiteley. She had been one of the earliest women students and was second-in-command to Prof. J.F. Thorpe in the Organic Chemistry Department. Her standards were tremendously high and woe betide anyone who fell below them either in work or in any other way. I am afraid we thought "Martha" as she was known to us (but not to her face) very old-fashioned, but in years to come we realized how much we owed to her. Lady Holland, the wife of the Rector and Mrs. (afterward Lady) Thorpe, Mrs. Philip and Mrs. Baker, wives of Professors, took a great interest in the Association and were Honorary Members.

I do not think we considered ourselves as second class citizens in a men's College, probably we thought of ourselves as superior! We had few amenities and we were not allowed into the Bar of the Union, but this was just a fact of life. All, male and female, called each other by our surnames. Generally we had no idea what people's christian names were unless it was to distinguish between people of the same family such as the three Conradys, Hilda, Doris and Rene or between people of the same surname. To this day some of us call each other by our surnames, much to the amusement of the younger generation.

We played various games. Tennis was played on the courts, now gone, outside the Union. One of my friends writes "We played one or two Women's Colleges. I played with Lewkowitsch, which suited me, as she did all the work and I wasn't good". Anyone knowing Elsa Lewkowitsch, now a member of the RCSA Committee, and her amazing energy can well believe that. We had a table-tennis (which I am sure we called ping-pong) team which had matches with an I.C. men's team and with other Women's Colleges. We played amongst ourselves on a large table in the RCS Common Room. There was some arrangement by which we belonged to some London University Women's Organisation and were allowed to row on Regent's Park lake in Bedford College boats. I remember much hilarity but little skill. T.M. Holmes (Mrs. Bailey) remembers going to meetings when she was President in 1925 and getting very cross because the Chairman used to talk about "the small Colleges like Maria Gray and Imperial College" because we had so few women.

A number of social functions were held, such as the Annual Dinner in the Union and the Hallow E'en Party in the Women's Common Room on the ground floor on the East Corner of the Union. For some reason the outer door to it was locked early in the evening after which time access to and from it was only through the men's lounge. According to *The Phoenix* for November 1924, the party was held on the 5th November that year and it is reported that *"the accomodation was trained to breaking point there being 30 members present .. After an excellent tea in Committee Room A members repaired to the Common Room and celebrated the 5th with indoor fireworks, a Freshers' entertainment, charades and dancing to the gramophone".* Needless to say a number of us were members of the Dram. Soc. and we had at least one theatre party each term. One such was to "Yellow Sands" and we sat in a very steep gallary and some of us ate oranges much to the disgust of others.

One of these expeditions was preceded by supper in the RCS Common Room and I still have a photograph of us sitting round the table.

Mention of the RCS Common Room leads me to what is the most vivid recollection I have of my time as President, it's refurbishing. It was at the top of the main stairs of the RCS on the East side. There were three interconnecting rooms the last being the only Ladies' lavatory in the building, (oh those stairs!) It contained besides the loos and washbasins an enormous table, a minute mirror, a row of coat-hooks and a gas ring. The only thing I remember of my first day at College is of another Fresher, Scott-Moncrieff, telling me we were allowed to make tea on it! The middle room was the Common Room and had another enormous table, an old sofa and some hard chairs, and the first contained nothing but cupboards. What was kept in the cupboards I can't remember but I know

Extract from the RCS Handbook 1974–1975, pp. 34–35.

we had nowhere to keep any personal belongings. Lady Holland, Mrs. Thorpe and Dr. Whiteley decided something must be done to make the place more comfortable. The first thing to do was to in some way to cover the hideous white tiles in which the whole of the college was lined from floor to ceiling. The Clerk of the Works of the College was "Jimmy" Peacock, known to latter students in a different guise, and he said that he had had the tiles most successfully painted in another room and invited me to come and look at it. The room turned out to be the men's Staff lavatory! Students today cannot imagine the embarrassment such a visit caused to a woman in those days. I was terrified one of the Staff would come in and I have never forgotten it. Anyway the tiles were painted, but there is a difference of opinion as to what colour. I think it was chocolate! Others think sunshine or mustard and some chocolate and cream. We threw out the old furniture and with great help from our Honorary members we collected enough money to buy armchairs, small tables, a bookcase and two pictures. The pictures were Cecil Aldin prints chosen by Mrs. Thorpe and myself and I thought spendid. I was somewhat dashed when a member said, "Gracious, it would have been better to have had railway posters". She still thinks the same for she writes" . . . wishy-washy-all rather alike, hunting, coaching, ye olde nonsense in the Harrods style". Two of the lecturers (male) P.C. Bull and D.A. Spencer each gave us one of their own paintings which probably helped to counteract the Establishment effect. Lady Holland gave us a tea service and the College two (not new) fair-sized tables to take the place of the very large one and a rather more comfortable sofa, and two rugs. One member remembers going to a furniture store in Oxford Street with Dr. Whitely, whose father had been Managing Director or something, and she felt like a lady-in-waiting on the Queen, with doors being flung open and the greatest respect paid by bowing officials. They were helped to choose some chintz material for loose covers by one of the high-ups and got it at wholesale price. A little seamstress from the Paddington area came and made and fitted the covers for 7/6 a piece and this included the bus fare! After the last war the then President wrote and asked me if I had any evidence to prove that the furnishings belonged to the Association as the College was turning the rooms into labs. and were claiming the furniture etc. I had no evidence, all I could say that I *knew* that it was so. I never heard what happened, but I can't think modern students would appreciate those pictures.

Recently I tried to see the old haunts but the RCS as I knew it has gone. Nothing remains of the Chemistry side but an aching void, but overlooking it are three windows which could have been the Common Room ones, but from all appearances the only way to reach them is by helicopter on to the roof. Sadly I turned away and went across to what used to be Imperial Institute Road to visit Mrs. Pingree, the IC Archivist. I am most grateful to her for all the trouble she took to look out copies of the Phoenix and the other documents for me and the fascinating selection of photographs. If anyone wants to be transported back to the old days I can thoroughly recommend a visit to the Archives.

"Gosset"
7th May 1974.

I.C.W.A. Today

On a certain day in February, a drastic increase in humidity occurs in the vicinity of the Beit Quadrangle. The event is the annual Pancake Race between I.C. Union Executive (of which our illustrious RCS President is a member) and the ICWA Executive. One and all have the dubious privilege of testing the principles of fluid dynamics which they are supposed to have learned within the hallowed walls of Imperial College and at the same time practice their ability at fire fighting with non-existent fires. This, however, is not the normal guise in which ICWA may be found during the rest of the year.

Unless you happen to be one of the Oneintenpointfive at I.C. or have the most exquisite and extrovert kneecaps in this fine academic institution and thus have qualified as Mr. ICWA, then it is really rather improbable that you are a member of Imperial College Women's Association. ICWA does, however, most certainly not exist to provide tea in great abundance for its own members, but rather to improve the social life of ALL students at college, and this is precisely what it does.

In each department of RCS there is a friendly and approachable ICWA rep just waiting to persuade you to attend what will indubitably be (if and only if graced by your company) the next "good thing" organised by ICWA during the year, just some of the events which will occur are the Halloween Folk Dance, lunch parties (which could never be worse than eating a Mooney) and one of the events of the I.C. calender, namely the Easter Ball. The first big event in the Autumn term is the ICWA Freshers' Dinner (on Monday 7th October) at which all staff and students and, of course, particularly lady freshers are most welcome. After this dinner, Mr. ICWA is elected. Last year, this honour (which for many years had been held by a minesman) was bestowed upon Digby James, co-editor of this publication, for his beautiful palpitating patellas. Perchance the gentlemen of RCS will see fit to establish a new tradition. See you there.

MARY SHORT

A 1963 flyer advertising the tradition of the Shrove Tuesday Pancake Race which clearly continued, as it is was referred to in the 1974 handbook text.

ICWA — Imperial College Women's Association this was an early precursor of the scientific societies and had been known as **ICWSA** — Imperial College Women Student's Association.

Imperial College Women's Association, 1913–1914 — Seated third from left; Martha Annie Whiteley, first President.

The President, Miss Theodora Mary Holmes, Women's Association, *Phoenix* 1926.

Imperial College Women's Association Tea party, 1925.

A joint venture between WSET and ICWC. The Women's Club continues to host the termly lunches.

Tie Clubs and Women

Toni Semmence

There are various clubs and societies that students can join associated with the Students' Union, but there are also those known as Tie Clubs. Members were obliged to wear a tie on a certain day each week or accept a forfeit. For many years these were men only clubs, but now the '22 Club has formally instituted a women's branch, called the XV Club, after the year it was founded 2015. Toni Semmence, a Postgraduate student, the initiator of the Club, is the first President and writes:

As President of the newly founded XV Club (the "Fifteen Club") at Imperial College London I am writing to introduce the club to you.

Toni Semmence 1st President of 1st Women's Tie Club.

The XV Club is a fledgling women's social club, modelled closely on the traditional 'tie clubs' that have been around Imperial for nearly 100 years, but with a womanly slant. You have no doubt come across The '22 Club, The Links Club and The Chaps Club, associated with the Royal College of Science (RCS), City and Guilds College and Royal School of Mines respectively. You may also recall that, in the main, women have not been eligible to join these old-established clubs.

The XV Club is closely associated with RCS and has been formed with support and advice from the '22 Club, and most of our members will come from RCS departments. But honorary members from outside of the RCS are also welcomed.

Our objective, as described in our rulebook is to "promote good fellowship among those past and present [female] students of the RCS, who have participated actively in the life of the College, and who can be relied upon to uphold the prestige and further the interests of the College." Less formally speaking, we aim to create a close network of current students and alumnae and to arrange social events where we can meet and enjoy each other's company, and of course that is extended to our honorary member too.

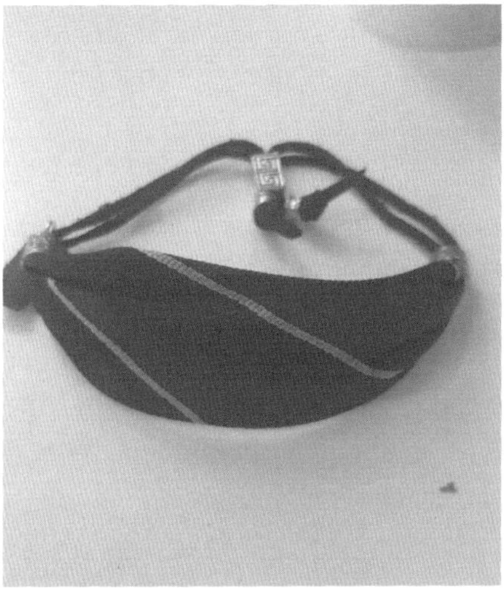

A bracelet in RCS colours replaces the tie.

In the XV Club, we already have a vibrant community of College members, and an enthusiastic executive committee consisting of present students and past students of various vintages. Together we have enjoyed several initial get-togethers, both girls-only and also mixed events in conjunction with the '22 Club and the other tie-clubs. We are planning an exciting social calendar for this year including formal dinners and a variety of less formal events, in the Union Bar and at other venues around London.

In 2016, the Chaps Club initiated their first female member and the Links initiated their first for a few years. For 2016–2017, both the Links Club, and the Chaps Club, include women on their committees.

Imperial College Women Students Union Presidents and Deputy Presidents, Secretaries and *Felix* Editors

Presidents

1970–1971 Judith Walker (PG Materials Science 1970) 64th SU President.

1978–1979 Mary Attenborough, (Mathematics PhD 1982) 73rd SU President.

1983–1984 Gaynor Lewis (BSc Mineral Resources Engineering 1983) 78th SU President.

Christine Taig (Mechanical Engineering 1986) 81st SU President.

1991–1992 Zoe Hellinger (Computing Science 1991) 86th SU President.

1994–1995 Lucy Chothia (Chemistry 1994) 89th SU President.

1995–1996 Sarah White 90th SU President (London University SU President 1996–1997).

1999–2000 Natasha Newton (Medicine 2002) 95th SU President.

2005–2006 Sameena Misbahuddin, 101st SU President.

2008–2009 Jennifer Morgan, (Biology 2008) 104th SU President (RCSU President 2007–2008).

2015–2016 Lucinda Sandon-Allum (Biology 2015) 111th SU President.

Deputy Presidents

1980-1981 Rachel Snee (Microbiology 1980).

1981–1982 jointly Christine Teller and B.J. McCabe.

1983–1984 Christine Teller (Physics 1984).

1986–1987 Jackie Peirce (Physics 1986).

Deputy President — Clubs and Societies

1996–1997 Sarah Corneille (Biology 1996).

1998–1999 Marie Nicholaou (Computing 2001).

2002–2003 Nona Ahamat (Biochemistry 2002).

2008–2009 Lily Topham (Medicine 2011).

2009–2010 Jennifer Wilson (Biochemistry 2011).

2010–2011 Heather Jones (Physics 2012).

2011–2012 Monya Zard (Physics 2011).

2013–2014 Yasmin Edwards (Biology 2013).

2014–2015 Abigail de Bruin (Chemistry 2014).

Deputy President — Graduate Students

2006–2007

Shama Rahman (Physics PhD 2014).

Deputy — Education

2003–2004 Katherine McGinn (Medicine 2008).

2005–2006 Sarah Khatib (Biochemistry 2005).

2007–2008 Kirsty M. Patterson (Geophysics 2010).

2008–2009 Hannah Theodorou (Medicine 2011).

Education

2013–2014 Natalie Kempston (Biomedical Sciences 2012, Public Health Masters 2014).

Deputy Presidents — Finance and Services

2004–2005 Sameena Misbahuddin (Computer Engineering 2004).

2016–2017 Rachel Blythe (Geophysics 2016).

Welfare

2010–2011 Charlotte Ivison (Ecology and Environmental Biology 2012, Environmental Technology 2015).

2012–2013 Rebecca Lane (Physics 2012).

2013–2014 Marissa Lewis (Medicine).

2015–2016 Jennie Watson (Medicine).

2016–2017 Emily-Jane Cramphorn (Biology 2016).

Honorary Secretaries

1968–1969 Miss J.E. Pearson (Chemistry 1969).

1974–1975 Jacqui Gerrard (Mathematics 1976).

1980–1981 Miss E.A. Lindsay (Chemistry 1980).

1982–1983 Jointly Mary Freeman (A.D. Grimshaw) (Computing Science 1982).

Honorary Secretaries — Events

1989–1990 Fiona Nicholas (Biochemistry 1989).

Felix Editors (Student's weekly magazine)[i]

2014–2015 Philippa Skett (Biology 2013, Science Communication 2015).

2015–2016 Grace Rahman (Biology 2015).

President's Address

Lucinda Sandon-Allum — Women@Imperial Week 2016

I'm now coming to the end of my tenure as President of the students' union here at Imperial — it has been an amazing year, and a privilege to be part of the team of elected Officer Trustees.

Imperial College Union is a proudly international institution, supporting student groups of over 100 nationalities and organising student activities that take place all over the world. Following last week's referendum result, the Union will be working closely with the College to establish what impact the UK's withdrawal from the EU will have on our members and how we can best support them.

It is testament to the strong relationship between the Union and the College that we often work together on initiatives which are important to our community. One of the things that most surprised me when I started in this role was the genuine dedication of College staff to students' education and university experience. I was expecting to have to fight to be heard in meetings, but it was quite the contrary. Staff were always keen to hear a student perspective and to consider the impact on students when making decisions.

Initiatives such as the Women@Imperial and developments in the College's accommodation offer have benefited from this collaborative approach.

The Student Academic Choice Awards recognising staff for their contributions to education and the student experience at the College are our way of acknowledging this dedication from staff at the College, and a chance to thank the staff that have made a real difference to students. Students submitted over 600 nominations for the awards this year and from the number of staff who attended the presentation event, it was clear that the recognition was valued. Many congratulations to all of the winners from the staff community.

I know that the new team of Officer Trustees, starting on 1st August 2016, will want to continue working alongside the College on key issues for students. Thank you to all of you who have supported us this year, it has been a pleasure working with you and I know you will extend a warm welcome to the new team later in the summer.

Best wishes,

Lucinda Sandon-Allum President, Imperial College Union

Endnotes

[i] *Felix* is the weekly student newspaper of Imperial College London and covers a wide range of journalism relevant to the students. We print about 30 issues a year, which are distributed every Friday. As well as the usual sections such a news, comment, careers and sport, the paper also includes sections which cover film, music, the arts and local events, as well as light-hearted sections such as the centrefold and hangman. Anybody can get involved, no experience is needed and you will be made to feel welcome right from the start. As well as producing a weekly newspaper, magazines are also produced from the office, which include the science magazine *I, Science* and the arts magazine *Phoenix*. *Felix* is run by a full-time sabbatical but the content is provided by the students. Retrieved from https://www.imperialcollegeunion.org/activities/a-to-z/381. Retrieved from http://www.union.ic.ac.uk/media/felix.

Recruiting Women Students — Publicity

Imperial College has always been aware of the need to continually recruit students, and that to appeal to women it was necessary to encourage them to look beyond the accepted wisdom, that science and engineering were male bastions.

> The then Registrar, Peter Mee, worked consistently with external organisations and with Imperial colleagues to further the cause of women's recruitment. In 1972, he wrote to the President of the Women's Engineering Society: we are, not unnaturally at Imperial College, very keen on increasing the number of girls who come here to read science and engineering degrees. With this in mind we are planning to hold a conference on 26/27th September of College Staff and of teaching staff from girls schools and mixed schools from all over the country...We also hope to have present a number of current women students and a number of our past women students who are now embarked on successful careers…[i]

He continues that there seems to be an understanding at girls schools that science is an acceptable career but a misunderstanding about the appropriateness of both degree courses in engineering, and the possibility of subsequent suitable careers.

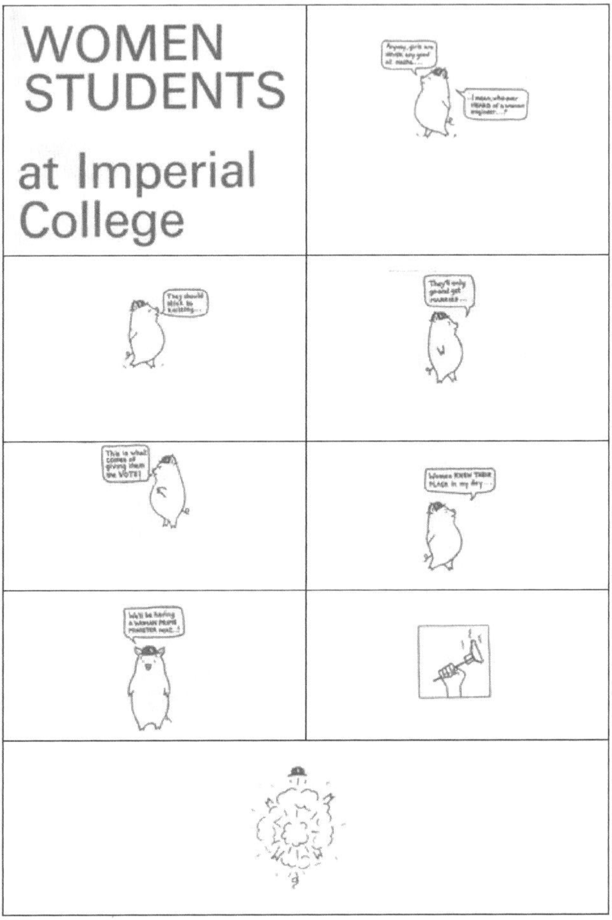

Other attempts were made to encourage women to study at Imperial in the 1950s, to the 1960s, 1970s and 1980s, but being in the predigital age, this often took the form of leaflets, some more exciting or innovative than others.

Blocks of text were acceptable forms of outreach, but even so were rather stultified. A 1968 leaflet sought to reassure women that they would be cared and catered for as women at Imperial. It stated that there were women's societies, a good social life and welfare services and non-scientific activities such as the recreational library and General Studies Lectures.

The residential hostel was adjacent to the Students Union and therefore presumably safely within the College boundaries. This was clearly seen as a bonus as was the increase in number of places to nearly 50! The total number of women students was then around 221 and total student numbers, men and women, was then around 3,400.

A later snappier glossy version of this title was produced in 1975, the era of the concept and phrase "Male Chauvinist Pig" (MCP) to denote an arrogant male, dismissive of the qualities, potential and attainments of women. A leaflet using text and ironic cartoons capitalising on this was produced.

Examine Imperial's website to verify the responses to the MCP speech bubble texts that can be given today as follows:

MCP speech bubble texts and responses
Anyway, girls are NEVER any good at maths…

Maths: Oh yes they are very good at it: A* or A at A Level minimum for undergraduate entry depending on the course.
…I mean who ever heard of a woman engineer…?

Engineer: Imperial College has, we have trained and employ many women engineers from the 19th to the 21st centuries.
They should stick to knitting…

Knitting: Yes knitting bones, cells and materials together as in Materials, Bioengineering, Medicine.
They'll only go and get MARRIED…

Marriage: Yes they marry, and also bring up families, and also have scientific careers and Imperial College supports family life.
This is what comes of giving them the VOTE!

Enfranchising women: Yes, promoting equality in all areas including currently heading Imperial College: **The First woman President of Imperial College was appointed in 2014.** There are now women in all roles and at all levels: academic, administrative, professional, technical and support services.
Women KNEW THEIR PLACE in my day…

Knowing their place: Yes, and it is in science, technology, engineering, medicine and business.
We'll be having a WOMAN PRIME MINISTER next..!

Woman Prime Minister: Yes, and she was a Chemist: Baroness Margaret Thatcher (1925–2013) Prime Minister UK 1979–1990.
The leaflet *Women Into Science and Engineering* featured the MCP

Is there no limit to what they can do? No, there isn't, as we know from the womens' careers that have begun here as students, the development of their careers in all spheres of activity, some that are forged at Imperial, and those that continue beyond.

Endnote

[i] Peter Mee Registrar Imperial College to Miss P.L. Hedges President, The Women's Engineering Society 2nd August 1972 Archives Imperial College.

Christine Teller – Physics

Imperial College was an easy choice for me. I liked the atmosphere and was keen to live in London for its theatre and concert facilities. The unequal male/female ratio didn't bother me and I found women well accepted in the Physics Department.

One of the greatest advantages of taking Physics at Imperial is that the course, whilst being broad enough to allow almost any diversification after graduation, is also regarded as a sound and useful education by employers.

In my first year I lived in a residence close to the College. This was a good place to start a social life and I also made friends quickly through one of the student societies.

After my first degree I did an MSc in Optics at Imperial and spent a further year working full-time as Deputy President of the Student's Union.

When I left the College I started working in management consultancy. I don't use my physics training directly, but I use my analytical and technical skills and find the administrative experience of working in the Union has given me a headstart in my job.

Gaynor Lewis – Mineral Resources Engineering

I chose Imperial because of its good reputation and because it offered a comprehensive general engineering course which appealed to me as a solid grounding for a variety of interesting careers. I found the practical work on the course especially stimulating and valuable, making a good link between the theory and the everyday reality of the mineral industry.

My first Easter vacation was spent on a geological field trip and the next at the College's mine in Cornwall on a surveying course. During the summer vacations I worked overseas at diamond and tin research institutes.

In my final year, I applied for several jobs in the mineral industry and was offered three. However, at the same time I was elected President of the Students' Union and so I decided to spend another year at the College working on behalf of students. This experience encouraged me to apply for managerial posts with the big banks in London, whose responses showed that they were impressed with both my academic and my administrative experience at Imperial.

Anna Gigli – Statistics

I decided to study maths when I was about six, I think, so no-one in my family was surprised when I went on to study it at university. When I tell people that I did maths they are usually a bit surprised, but because they think it is a hard subject, not because I'm a woman.

After I finished my first degree in Rome I worked for a year as a researcher; then I got a scholarship to come to Imperial to do an MSc. I already knew London before I arrived and was attracted by the large number of cultural and scientific events taking place. The College itself was well known to me for its reputation in the field of statistics.

When I arrived, I found everything more exciting than I'd expected. The social life here is well-organised and I found it easy to settle in. I work hard at my studies but still find time to take part in many social activities. The imbalance of sexes doesn't matter to me. About half the students on my course are women and so are a lot of the lecturers.

Barbara Moont – Mechanical Engineering

When I was at Imperial College in the 1960s, the women in the Mechanical Engineering Department could be counted on the fingers of one hand. Having come from an all-girls school, I found this a bit strange at first but I soon got used to it.

I enjoyed College life living in hall and despite a few early doubts felt that I had made a wise choice of subject.

After graduating, I went to America to do a Master's degree. In all, I was there for two years researching into nitric oxide emissions from car engines and working at the instrumentation laboratory where the guidance and navigation for the Apollo space program was done.

Back in England I joined ICI as their first woman mechanical engineer and then worked in the Biomechanics Unit at Imperial College researching into the mechanical properties of the human spine.

Since having a family, I have worked from home as an energy consultant and now teach physics at a secondary school. I have also started my own company.

Later leaflets in the 1980s featured women speaking for themselves.

Centenary Prizes in the Name of Imperial Women Established 2007

Faculty of Natural Sciences

Physics
Anne Thorne

Chemistry
Martha Annie Whiteley
Frances Gore Micklethwaite

Mathematics
Yael Naim Dowker

Faculty of Engineering

Aeronautics
Elaine Austin

Bioengineering
Stella Bagrit

Civil Engineering
Letitia Chitty

Chemical Engineering
Julia Higgins
Mechanical Engineering
Margaret Fishenden

Computing
Ada Lovelace

Earth Sciences /Geology
Janet Watson

Electrical Engineering
Hertha Ayrton

Materials
Constance Fligg Tipper

Faculty of Medicine

Brenda Ryman
Julia Polak
Rosa Beddington
Claire Sharrock
Sheila Adam

In Imperial's Centenary year, women were celebrated by establishing Faculty and Departmental Prizes in the names of women who made an outstanding contribution to Imperial.[i]

Name

Yael Naim Dowker (1919–2016).
First Woman Reader in Mathematics Department at Imperial College.

Specialism

Ergodic theory, Measure theory.

Joined Imperial

1954.

Education

John Hopkins University; Radcliffe Harvard 1948.

Career Roles

Staff Member Radiation Laboratory MIT 1944–1945; Emmy Noether Memorial Fellow Institute for Advanced Study Princeton 1948–1949; Assistant Lecturer Manchester University 1950–1952; Assistant Lecturer Westfield College London 1952–1954; Lecturer 1954, Reader 1966–1985 Mathematics Department Imperial College; Sabbatical Research and teaching Alberta University 1970–1971.

Award in Her Name

Yael Naim Dowker Centenary Prize Imperial College.

About

Born in Tel Aviv, Yael undertook degree level study in the USA. She came to Imperial from Harvard highly recommended for her fundamental work in Pure Mathematics, particularly Ergodic theory, and published in significant journals from 1947. She was a collaborator of the prolific number theorist, Paul Erdos. Her most famous student was William (Bill, later professor) Parry FRS, who took a PhD with Yael, in her areas of expertise. Yael married Clifford Hugh Dowker in 1944, whom she met whilst he was an instructor at John Hopkins University. In 1951 he followed her to England without having a job himself, and later became Reader, then Professor in Applied Mathematics at Birkbeck College. Yael and Clifford had a daughter born in 1960. They worked with gifted children suffering difficulties at school, for the National Association for Gifted Children.

Name

Lady Stella Bagrit

(d.2001) founded The Sir Leon Bagrit Memorial Trust in honour of Sir Leon (1902–1979). Lady Bagrit was instrumental in the Trust's support for the creation of the Sir Leon Bagrit Centre in 1991, later, the Bioengineering Department. MSc bioengineering scholarship schemes, and student achievement awards continue to be supported by the Trust. The annual Bagrit Lecture commemorates the Bagrit name. The Stella Bagrit Centenary Memorial Prize commemorates Lady Bagrit's huge involvement with bioengineering at Imperial.

Name

Brenda Edith Ryman (1922–1983)
BA, MA, PhD.
First Married Mistress of Girton College Cambridge.

Specialism

Biochemistry, liposomes, innovative medical Education
Joined Imperial 1972.

Education

Girton College Cambridge 1941–1943; Birmingham University 1946–1948.

Career Roles

Research Chemist, Glaxo; Assistant Lecturer and Reader, Royal Free Hospital Medical School; Chair, Biochemistry Charing Cross Hospital Medical School; 1976–1983 Mistress, Girton College Cambridge.

Career Awards

Brenda Ryman Research Fund Charing Cross Hospital; The Brenda E. Ryman Centenary Memorial Prize Imperial College.

Brenda Ryman was dynamic and inspiring in her research, her teaching and in the development of academic departments as centres of excellence. She led Girton College to modernise and to admit men.

Endnote

[i] Exceptions to this are those women not part of Imperial, but who have made outstanding contributions in their subject e.g. as for Ada Lovelace Computing, and in their subjects at another institution with colleagues at Imperial, as in the case of Rosa Beddington, Claire Sharrock, Sheila Adam.

Women@Imperial:

Gender Equality, Athena Swan and Women Ambassadors, Scholarships, Scientific and Entrepreneurial Initiatives at Imperial and for Returners to Scientific Careers, Women@Imperial Week

Imperial takes gender equality very seriously. Major initiatives have been taken and strides in equality have been made over the last few years, building on a progression from very few women being at Imperial to over 3,600 women staff, plus 5,897 women students.

The Athena SWAN Charter was established in 2005 to encourage and recognise commitment to advancing the careers of women in science, technology, engineering, maths and medicine (STEMM) employment in higher education and research. Imperial College was a founder member of the Athena Swan Charter.[i]

Previously, the Academic Opportunities Committee oversaw Imperial's progression with Athena SWAN. In 2013, the Committee was restructured into the Academic Gender Strategy Committee (ASGC), which champions and oversees the advancement of gender equality at the institutional level, and The Athena Committee which is operationally-focused. Its main purpose is to ensure that changes are made to departmental practices and culture so that departments gain, or retain, Athena SWAN awards, increasing the award grade level where possible.

The Athena Committee is chaired by Professor Dot Griffiths, and membership includes the Faculty Ambassadors for Women, the College Athena SWAN Coordinator, and a representative from each department

The AGSC, chaired by the Provost, champions and oversees the advancement of gender equality at the institutional level, providing senior level support for Athena SWAN activity.

Caroline Fox first woman Director HR, Programme Manager of the Athena Project 2001–2007.[ii]

The annual Athena Lecture was established in 2001 by the Academic Opportunities Committee to celebrate the achievements of women in science, engineering, technology and medicine.

The Lecture has been given five times by then members of staff, Maggie Dallman (when Principal of the Faculty of Natural Sciences), Dame Julia Stretton Higgins, (when Principal Faculty of Engineering) Julia Buckingham, (when Pro-Rector (Education) Sue Ion (Alumna and Visiting Professor) and in 2016, Imperial President Alice Gast.

Imperial was awarded a Bronze Athena SWAN institutional award in 2009. In November 2012, Imperial successfully submitted for Silver, becoming one of a very small number of universities to hold a Silver award.

All departments at Imperial are expected to eventually hold an Athena SWAN award. Currently 15 departments have awards, ranging from Bronze to Gold, as detailed below, and the other departments are working on applications.

Faculty of Engineering Departments

Aeronautics: **Bronze**, November 2014.
Bioengineering: **Bronze**, April 2015.
Chemical Engineering **Silver**, renewed April 2014.

Civil and Environmental Engineering **Bronze** application planned for 2016.
Computing **Bronze**, renewed November 2013.
Earth Science Engineering **Silver** April 2012.
Mechanical Engineering **Bronze** application planned for 2016.
Materials **Silver**, renewed November 2013.
Mechanical Engineering **Bronze** application planned for November 2015.
Dyson School of Engineering new Department.

Faculty of Natural Sciences Departments

Centre for Environmental Policy **Bronze**, Application Planned.
Chemistry **Gold**, April 2013 (10 academic women).

Sue Gibson Professor of Chemistry, CChem, FRSC, OBE 2013 for 'services to chemistry and science education; Director of the Graduate School at Imperial College. First holder Royal Society's Rosalind Franklin Award, Sue funded a series of UK lectureships for internationally renowned female chemists.[iii]

Life Sciences **Bronze**, renewed April 2015. Two Professors, two Readers, three Senior Lecturers, five lecturers, two Associate Provost (Academic Partnerships), 1 Research fellow Career Development, two Teaching Fellows, fourteen Research Associates/Fellows.

Mathematics **Bronze**, November 2012. One Professor, four Readers, one Principal Teaching Fellow.

Physics **Silver**, renewed 2015. Five women Professors, one Reader, four Senior Lecturers, two lecturers, one EPSRC Career Acceleration Fellow.[iv]

Faculty of Medicine Departments

Institute for Clinical Sciences **Bronze**, 2014.
Department of Medicine **Silver**, November 2014.
National Heart and Lung Institute **Silver**, renewed April 2014.
School of Public Health **Silver**, renewed November 2013.
Surgery and Cancer **Bronze**, April 2014.

Most Departments have Women's Committees and initiatives

Physics: **Departmental Juno Transparency and Opportunity Committee** was established in 2007, to implement the Institute of physics Juno Code of Practice, which aims to support best practice for all staff with particular attention to putting in place practical actions to address the under representation of women in university physics.

Women in Physics Student Society

'We're a group of enthusiastic young researchers who are passionate about physics and excited about a future in science. We know that for the best science to happen we need to access the talent of the whole of the workforce. We're committed to increasing the amount of women in the physics department at every stage of their careers: from undergraduate to professor. To do that, we organise events that celebrate women's achievements and talk about issues we might face'[v]

Athena Swan Development

In May 2015, the Athena SWAN Charter was expanded to recognise work undertaken in arts, humanities, social sciences, business and law (AHSSBL), and now includes staff in professional and support roles, as well as trans staff and students. The Charter now recognises work undertaken to address gender equality more broadly, and not just barriers to progression that affect women.

Through its award system, Athena SWAN aims to recognise and celebrate good practice on recruiting, retaining, and promoting women in higher education. Athena SWAN also provides a valuable framework for self-assessment and action planning.

Dr Liz Elvidge Head, Postdoc Development Centre, is responsible for the support and development of Imperial's 2,300 researchers and fellows. She is an Athena Swan assessor and a member of the cross-college Athena Swan committee as well as a number of departments Swan Committees. She has two step-children and a granddaughter. She is the sole carer for her husband, who recently took early retirement due to ill health.

Jackie King Photographer

Dr Liz Elvidge, Head of the Postdoc Development Centre. Portrait taken as part of the Women@ Imperial exhibition held on 13 March 2015.

The Women Faculty Ambassadors

This scheme is associated with success in Imperial gaining an Athena SWAN silver award. The award recognises the College's demonstrated commitment to recruit, retain and advance women academics. The Ambassadors scheme was established in October 2007 to promote this message and encourage participation in the scheme.

The Ambassadors will support fellow female academics in their professional development and recommend tutoring or mentoring facilities if they are needed. In addition, they will work with the Principals and Deputy Principals on annual events and activities for women in the faculties. They are selected by their faculties and hold the posts for three years. The first three were appointed in 2008

Professor Maria Petrou, Department of Electrical and Electronic Engineering. Maria's advice: "Do not give up! Persevere! Never put yourself down even as a joke! If you say jokingly "Oh, I have no idea about this!" — "this" being something bang in your area of expertise — believe me, there will be people who will take you seriously and believe that you really have no idea. Never undersell yourself!"

Professor Maria Belvisi National Heart and Lung Institute.
Maria's advice: "I think the most valuable piece of advice is to learn as much as you can from a positive role model. Also stay focused on your career objectives even during the bad times."

Professor Lesley Cohen from the Department of Physics.

Lesley's Advice: "Participate in this process. It is a wonderful opportunity to help to make positive change. Your voice and your experience really matters. I would be delighted to hear your views.
2016 Ambassadors

Professor Ann Muggeridge Petroleum Engineer, Faculty of Engineering, Earth Sciences Worked in Industry Joined Imperial 1995. She worked part time from the birth of her daughter in 1998 until 2008. She now works full time but continues to work flexibly, often from home.

Karen Makuch Lecturer, Environmental Law & Energy Law, Faculty of Natural Sciences, Centre for Environmental Policy. "I aim to support fellow female academics in their professional development and provide opportunities for women across the Faculty to meet and support each other through networking and mentoring events.

Dr Victoria Salem NIHR Clinical Lecturer in the Division of Diabetes, Endocrinology and Metabolism, Faculty of Medicine Department of Medicine. As Athena SWAN Coordinator for the Faculty of Medicine, she supports and advises departments on their Athena SWAN work.

Fellowship and Scholarship Programmes

Innovative support includes The Stevenson Fund awarded annually, three bursaries of up to a value of £7,500 each. These are open to female students from Imperial in physics, chemistry, maths, life sciences and geology to undertake an international research placement of up to a maximum of three months with a leading female scientist in a similar institution. The Review Panel 2016 Professor Sue Gibson (Chair) Professor Maggie Dallman, Professor Dot Griffiths, Professor Joanna Haigh, Professor Ann Muggeridge, Lily Le Wong, PGR Life Sciences, 2015 recipient, Nikita Kamraj, UG Physics, 2015 recipient.

Entrepreneurs

The Althea–Imperial Programme was first launched in October 2014 by Imperial's senior leadership and founder of The Althea Foundation Alexsis de Raadt-St James.[vi] The Althea–Imperial Prize is designed to encourage a new generation of women innovators and entrepreneurs. It is a personal and professional development programme aimed at women studying at Imperial. It consists of a series of workshops designed to help you develop your enterprising idea. The programme will include high-profile speakers and tailored one-to-one mentoring.

Once the programme is complete, all participants can opt to submit their ideas to the Althea–Imperial Prize. Both, the programme and its associated prize have been designed to encourage female students, through a series of exclusive development and mentoring opportunities. As individuals or by leading a team, these students will see a pioneering project through to fruition.

The Business School[vii]

Women in Leadership is dedicated to the success of women at Imperial College Business School by providing meaningful experiences to students that foster personal and professional development. The scheme seeks to attract the world's best female students, and to provide them with the skills and network to support a successful career in business. At Imperial, we are committed to gender parity on all of our programmes because we know that having more women in leadership roles will have a long-term positive impact on business.

Forté Foundation Fellowships

Imperial College Business School is a proud member of the Forté Foundation, a non-profit organisation that champions women's progression within business through access to education. We are pleased to offer Forté Fellowships to outstanding female students joining the programme.

The scholarships are awarded to women who demonstrate strong leadership skills through academic achievement, or in roles as team or community leaders, and show a commitment to helping women achieve their career potential through personal mentoring or community involvement.

Diane Morgan Associate Dean of Programmes and Forté Foundation Board Member.

Professor Dot Griffiths (see separate entry).

Dorothy Griffiths Scholarship, with these awards, we celebrate Dot's outstanding service and desire to push the boundaries within education and research, and most importantly to continue her ambition to attract the best and most brilliant minds to the Business School, shaping the future of the next generation of entrepreneurs, leaders and innovators.

Imperial Women's Scholarships

The scholarships are awarded to women who can demonstrate both depth and breadth of work experience as well as academic excellence.

Initiatives for Returners to Scientific Careers

The **Elsie Widdowson Fellowship Award** is an important component of the College's family-friendly policies and is open to eligible female and male staff. The purpose of the Award is to allow academic staff (Lecturer, Assistant Professor, Senior Lecturer, Associate Professor, Reader and Professor) to concentrate fully on their research work upon returning from maternity, adoption, surrogacy and/or shared parental leave.[viii]

A **Daphne Jackson Fellowship** is a unique fellowship designed to return STEM professionals to their careers after a break. Fellows normally carry out their research part time over two years, in a university or research establishment in the UK. The Fellowships are flexible and include a tailored training programme designed to update skills and knowledge and give support on return to research.[ix]

Imperial College's website is a rich source of information on the constantly developing and wide variety of scholarships and support initiatives Imperial can offer.

Celebrating Women at Imperial: Women's Week — Bringing Initiatives Together

Woman's Week, or Woman@Imperial, was instigated by Dot Griffiths in her role as Gender Envoy, in 2015.

It has become an annual event; a week of celebration of women and their achievements in all roles at Imperial: student, academic and support, professional and technical. It showcases women and their work and publicises the initiatives advancing Imperial as a good place to be.

Women's Week is coordinated by a group including Dot, and staff from Communications, Archives, HR, Athena, Outreach, Departments/

Faculties and Student Union representatives. Ideas engendered by the group and their colleagues — discussion is encouraged — are put into practice to make a week of events.

Photographs of women are taken, text about their work and work/life balance written, archival displays created, and the Main Entrance on Exhibition Road is filled with Women@Imperial branded display boards as a showcase for women and their work.

Events are held both in the Main Entrance and in departments. An evening is given to launching the week with talks and performances from student groups such as the *Imperielles*, the first all women *a capella* group at Imperial, and the ICU Belly Dancing Society. Women scientists discuss their research in accessible terms, giving demonstrations and drawing in visitors to participate at their demonstration stands. The pleasure and excitement all those involved reveal is a huge incentive to consider science as a career and to encourage a new generation of women to become involved in any of the many aspects of life at Imperial.

Faculties and departments are encouraged to take initiatives such as special lectures by women during the week, and to acknowledge International Women's Day as part of the W@I week. For example:

In 2016,

"The first Department of Life Sciences International Women's Day Lecture will be delivered by world leading scientist Professor Anne Dell"

In a lecture entitled *The sweet Things in Life,* Anne said:

In this talk I will discuss my life in science aiming to give glimpses into the exciting field of glycobiology with the help of some of my collaborative research.

Anne Dell CBE FRS, FMedSci[x]

Professor Carbohydrate Biochemistry, Division of Molecular Biosciences, Biopolymer Mass Spectrometry, Faculty of Natural Sciences.

Career

Chemistry University of Western Australia

PhD at the University of Cambridge supported by an 1851 Exhibition Research Scholarship.
Postdoc Imperial College rose through the ranks to a Personal Chair in 1991.
Head Biochemistry Department 1999–2001.
FRS 2002 and was awarded a CBE in recognition of her services to science in 2009.
Head of the Glycobiology Training, Research and Infrastructure Centre at Imperial College
Honorary DSc degrees from the University of Western Australia and the University of Waterloo Canada.

Professor Anne Dell.
Photographer Dave Guttridge The Photographic Unit[xi]

Endnotes

[i] In May 2015 the Athena SWAN Charter was expanded to recognise work undertaken in arts, humanities, social sciences, business and law (AHSSBL), and now includes staff in professional and support roles, as well as trans staff and students. The Charter now recognises work undertaken to address gender equality more broadly, and not just barriers to progression that affect women. Through its award system, Athena SWAN aims to recognise and celebrate good practice on recruiting, retaining, and promoting women in higher education. Athena SWAN also provides a valuable framework for self-assessment and action planning. http://www.imperial.ac.uk/equality/achievements/athena/.

[ii] At 2016 the Athena SWAN Coordinator/ Equalities Administrator, Human Resources Division is Mr. Rob Bell.

[iii] Jenifer Mizen, 175 Faces of Chemistry Royal Society of Chemistry. Retrieved from http://www.rsc.org/diversity/175-faces/. Accessed 26 July 2016.

[iv] Academic Women in Physics. Departmental website. Retrieved from http://www.imperial.ac.uk/natural-sciences/departments/physics/about/women-in-physics/academic-women/.

[v] Women in Physics Student Society. Retrieved from http://www.imperial.ac.uk/natural-sciences/departments/physics/about/women-in-physics/students/.

[vi] The Althea Foundation: "We fund people, ideas and initiatives that inspire us with their ingenuity, humanity and social impact. Our goal is to invest, meas-

ure and scale the expansion of important ideas. We provide both early-stage development funds and donations that are mission aligned with our objectives. We also partner and co-donate with other social impact investors, entrepreneurs, philanthropists and donors who share our goals' http://www.thealthea-foundation.org/index.html. Accessed 26 July 2016.

vii The Business School Imperial College Women in Leadership Scholarships for Women. Retrieved from http://wwwf.imperial.ac.uk/business-school/programmes/women-in-leadership/scholarships-for-women/.

viii Imperial College Elsie Widdowson Fellowship Award http://www.imperial.ac.uk/human-resources/procedures/family-friendly-policies/elsie-widdowson-fellowship-award/.

ix See The Daphne Jackson Trust website for details http://www.daphnejackson.org/

x Anne Dell was amongst the first investigators to successfully apply soft ionization mass spectrometry to carbohydrate containing biopolymers. Over the past thirty years she has led a major group that has been at the leading edge of the development of many mass spectrometric methods for glycomic and glycoproteomic analyses, especially in the area of biomedical research. Her laboratory provides structural underpinning for programmes of research seeking to define the biological roles that glycans play in health and disease. She is currently a Wellcome Trust Senior Investigator and heads the Glycobiology Training, Research and Infrastructure Centre at Imperial College. Retrieved from http://www3.imperial.ac.uk/newsandeventspggrp/imperialcollege/lifesciences/eventssummary/event_16-2-2016-10-2-20.

xi By Permission of Imperial College.

Industrial Sociology to Business School

Dorothy Enid Cole Wedderburn

Industrial Sociology to the Business School in Brief

1955 Mechanical Engineering Department Production Engineering Section established.
1965 Industrial Sociology Unit created which with Production Engineering became.
1971 Department of Management Science (DMS).
1969 Industrial Sociology Unit became independent from Mechanical Engineering.
1978 Industrial Sociology merged with Social and Economic Studies Department.
1987 Social and Economic Studies Department merged with Department of Management Science to become The Management School renamed:
2003 The Business School with Faculty status.
2008 Imperial College Business School.

Dorothy Enid Cole Wedderburn (1925–2012).
BA.

Specialism

Industrial Sociology.

Career Roles

Research Assistant, Senior Research Assistant, Board of Trade 1946–1950; Research Assistant, Senior Research Assistant Department of Applied Economics Cambridge University; Lecturer 1965, Reader 1970, Professor of Industrial Sociology Imperial College London 1977; Principal Bedford College London 1981–1985, Principal of the merged Royal Holloway and Bedford New College 1985–1990.

Career Awards

Honorary Fellow Ealing College of Higher Education 1985; Honorary Fellow Imperial College London 1986; Honorary Fellow Royal Holloway and Bedford New College 1991; Honorary DLitt University of Warwick, 1984; Honorary DLitt University of Loughborough 1989; Honorary Fellow Brunel University 1990; Honorary LLD Cambridge 1991; Honorary DSc City University 1991; Honorary DSoc Sc University of Southampton 1999; Honorary PhD London Guildhall University 2000.

Professional Affiliations

Royal Commission on the Distribution of Income and Wealth 1974–1978; Pay and Condition of Nurses 1974–1975; Social Science Research Council (SSRC) 1976–1982; Joint Science and Education Research Council and Social Science Research Council 1980–1982; Committee of Vice Chancellors and Principals 1988–1990, (this organisation, an official forum for HE institution heads, is, since 2000, Universities UK). Member Court of the University of London 1981–1990, Pro-Vice Chancellor of the University of London 1986–1988; Governing Board of London Guildhall University, (now merged into London Metropolitan University) 1989–1999; Loughborough University Council 1990–1993, Court of City University 1992–2000; Council of Goldsmiths' College London University 1993–2000 Prison Reform Trust Committee of Inquiry into Women's Imprisonment 1998–2000; Council member Advisory Conciliation and Arbitration Service (ACAS) 1976–1982; Trustee The Anglo-German Foundation for the Study of Industrial Society member Executive

Committee 1997–2004, 1987–2009; District Health Authority The Royal Borough of Kensington and Chelsea 1993–2002. Honorary President Fawcett Society 1986–2002.

Quote

Born

18 September 1925.

Died

20th September 2012.

Family Background

Father: Frederick Barnard, craftsman wood worker and trade unionist; Mother: Ethel, domestic servant; Siblings: one brother George Alfred Barnard (Mathematics Lecturer, Professor at Imperial College 1948–1965) one sister deceased as a child.

Marriage

First in 1947 to William Alan 'Max' Cole an economic historian divorced, 1960; second in 1962 to Kenneth William Wedderburn, (1927–2012) an academic Labour lawyer, later Lord Wedderburn of Charlton, dissolved late 1960s.

Education

Schooling

Walthamstow County High School for Girls.

University

Undergraduate

Girton College Economics Cambridge University, 1943–1946.

Degree

BA.

Awards

School Scholarship.

Text Discussing Career and other Comments

It was Dorothy's father who encouraged her to apply for university — he had the view that as her brother had the opportunity she should too. He was an active trades union member, and an influence, along with her brother, on her application to university and her lifelong social and left wing outlook and her views on equality.

Dorothy's obituarist in *The Guardian*[i] notes it was unusual for two children of such a working class family (their grandparents were blacksmiths on both sides) to attend university in the 1950s. The family clearly had some unique attributes.

During WWII, Dorothy's school was evacuated from London and although the women staff were close to their pupils, it was a dislocating experience for all concerned. Her experience at Cambridge University which followed was as a difficult place for a working class girl to be in 1943. Dorothy felt she arrived at Girton College from her ordinary, but evacuated school with none of the models of social mores that the girls from middle and upper class backgrounds had. She did however have the intellect and confidence in that, because she had won a scholarship to Cambridge, which sustained her mentally and physically. At Cambridge, she immediately joined the Communist Party, and she says the Party provided "a remarkable training for women"[ii] because there was no intellectual discrimination, and all members were encouraged to join in political debate and student events, and behave as leaders, so as to promote policies.

They required students to stick to their studies and show good organisational skills, which stood Dorothy in good stead throughout her career. She had never thought of an academic career for herself, but she got one almost by default. Although she left the Communist Party in the early 1960s, her left wing leanings maintained her in the intellectual and social company of such great figures as Eric Hobsbawm, the eminent historian.[iii] She remained interested in left wing issues such as the Campaign for Nuclear Disarmament (CND)[iv] which therefore led to science policy, which was relevant to her later post at Imperial College.

On leaving Girton in 1946, Dorothy worked at the Board of Trade (now Department for Business, Energy and Industrial Strategy)[v] in London, as a Research Officer, until 1950. She then returned to Cambridge on her first marriage, and took the great opportunity of working as a Research Assistant with Richard Stone (Nobel Prize Winner for Economic Sciences 1984)[vi] at the newly created Department of Applied Economics. The impetus for the Department came from the Cambridge economist John Maynard Keynes.[vii] The Department was an important post WWII development as it would lead research into aspects of British life in a way that was not conceivable prior to the great societal changes brought about by the war. British life needed thorough investigation so as to have data to work with in the development of a new order. During her time in the Department, Dorothy was involved in empirical research into subjects such as redundancy, class and the welfare state.

This was to be the first of newly created departments that she was recruited to. During her 15 years there, becoming a Senior Research Assistant and teaching undergraduates via the tutorial (or one to one teaching system), Dorothy realised that she *did* want an academic career.

She had married in 1947, a fellow undergraduate Max Cole, but the marriage ended in 1960. Dorothy married again in 1962 and she and her new husband Bill Wedderburn, moved to London because of his new job at The London School of Economics and Political Science (LSE).

Dorothy decided to study Sociology at the LSE, most particularly because she was disillusioned with economics and the move of the discipline towards mathematical modelling, away from human aspects. C.P. Snow's *Two Cultures* scenario (science vs. the arts) had held a fascination for her along with the influences of production systems upon management and workers. There was a natural progression in her research interests which led her to develop her distinguished career at Imperial College. In 1965 she joined Imperial in the second of the newly created departments of her career, The Industrial Sociology Unit. Dorothy was appointed Lecturer in Industrial Sociology,[viii] being promoted to Reader in 1970 and then Professor in 1977. Although she left Imperial in 1981, Dorothy remained a Senior Research Fellow until 2003.

Her initial move to Imperial was prompted by her interest in industrial organisation, therefore, working with Imperial's engineering departments, which was the original intention of her role, would enable her understanding to develop. Plus, she was keen to work with Joan Woodward,[ix] Professor of Sociology, who also had a practical interest in management.

Dorothy was Director of the Industrial Sociology Unit between 1973 and 1981, this role overlapped with her appointment as Head of the Social and Economic Studies Department (SES)[x] between 1978 and 1981 and then the Industrial Sociology Unit merged with the SES department.

The mid 1960s to the 1970s was a very exciting period for the development of the subject area at Imperial College with an inspiring team of women academics, including Joan Woodward and Sandra Dawson,[xi] Lisl Klein[xii] plus, a graduate student, Dot Griffiths,[xiii] who carried the mantel onwards to the present, and is now Emeritus Professor of Imperial College Business School. (See individual entries). What made the team so vibrant and attractive to students was their hands–on attitude to research, and their involvement in the real world of industry, industrial relations and intellectual socialism. Dorothy contributed to debates for, and with, the Trades Union Council (TUC),[xiv] on job enrichment and job satisfaction. She became an expert on white collar redundancy and healthcare of an aging population.

Dorothy wrote and published on age and the welfare state, redundancy and class, reform of justice as it affects women, poverty, industry and change planning. From the 1960s onwards, she contributed to new scholarly journals on the political left wing. She was well known to, and respected by, the intellectual socialist leaders of the 1960s and 1970s, and was seen by the editors who founded the annual journal the *Socialist Register* in 1963, as *the* person to write about the welfare state. Her article *Facts and Theories of the Welfare State* appeared in volume 2 in 1965. (Dorothy was also asked to comment on the piece describing the founding of the Register in 1995).[xv] She published in many other journals such as *The Sociological Review, New Society, International Social Science Review,* The *New Left Review, Journal of Social Policy and Socialist History Journal*

These journals epitomised the change in academic outlook of the period of the 1960s and 1970s. They provided a fresh view and expression for academics for the radically changing times. She also wrote for the more establishment titles such as that of the *Royal Statistical Society.* In 1971 she took part in a BBC documentary "How to be unemployed with dignity" and wrote a companion piece in the BBC journal *The Listener* entitled *Unemployment in the Seventies.*[xvi]

During the period 1969–1982, Dorothy held several appointments, outside her immediate role, one was a year's appointment as Visiting Professor at the Sloan School of Management at the Massachusetts Institute of Technology (MIT). She served on the committees of various funding council's. With her inside knowledge, Dorothy was influential in directing government policy towards retaining the funding councils and in keeping the needs of higher education on track at a time of swingeing cuts in funding and threatened restriction of courses, during the early 1980s. She had first-hand experience of dealing with these in her subsequent career development.

Her career then took a more administrative turn in Higher Education, as she became Principal of Bedford College in 1981, a post she held until 1985. Bedford College, part of the University of London was founded as College for women in 1849, but became coeducational in 1965. When she became Principal, it was clear that the College could not survive alone in the face of the government reductions in funding for higher education, and negotiations began with Royal Holloway College. The resulting merger in 1985, joined the Colleges under the name Royal Holloway and Bedford New College (RHBNC), now generally known as Royal Holloway, and she became its Principal until 1990.

There are two portraits of Dorothy in the art gallery of Royal Holloway College, one by Peter Greenham painted in 1985 and the second by Paul Brason painted in 1991, which show the esteem in which she was held by the Colleges she directed, despite the difficult decisions she had necessarily made to save the institutions. One decision was to close the expensive to run chemistry department, reduce staff numbers across the college and also sell off three of the most valuable paintings in the esteemed art collection of Royal Holloway. None of these decisions were taken lightly, but Dorothy was perhaps the best placed person to begin the negotiations, and in these circumstances she demonstrated her empirical approach to her work in labour relations and economics.

In the citation for Dorothy's Honorary Fellowship of RHBNC, the Chairman of the Governing Body from the merger period into the early 1990s, Lord Allen of Abbeydale, declared that it was important that her achievement was seen as that of a person of:

> great humanity and understanding of her fellow mortals, with no trace of arrogance and with an ability to laugh at herself — and also an ability to enjoy some of the day-to-day activities, such as cooking, which do not invariably go with high academic achievement.[xvii]

Dorothy wrote an article detailing the process of the merger in *Higher Education Quarterly* in 1991.[xviii] The Imperial College student paper *Felix* ran an article in October 1988[xix] suggesting that Imperial and RHBNC would merge. However, in 1989 it was reported this was not to be, but in a joint statement Imperial's Rector, Eric Ash, and for RHBNC, Dorothy Wedderburn, regretted the loss of opportunity for academic benefit but claimed that discussions had revealed "areas of collaboration which would be of substantial mutual advantage".[xx]

> Dame Sandra Dawson, Deputy Vice Chancellor of Cambridge University and KPMG Professor of Management, worked with Dorothy at Imperial and said: She had exceptionally strong values; a strong sense of the importance of education, and of being able to have shoes on your feet. She had a great sociological imagination, she was able to analyse something and create a narrative around it. Her work in the creation of industrial sociology was very important, and led to the setting-up of business schools. She was also great fun, she was very clear, very straight with the capacity to relate to anyone she met.[xxi]

Dorothy's ability to communicate across boundaries was recognised by Sir Walter Bodmer's request that she join his Royal Society Committee on the Public Understanding of Science[xxii] in 1985.

Dorothy's career illustrates that whatever she did, fed into the next stage of her career, that she did not limit her horizons and when opportunities arose she took them, even if she had not expected her career to go in the direction which they led.

Endnotes

[i] Retrieved from http://www.theguardian.com/education/2012/sep/21/dorothy-wedderburn Friday. Accessed 21 September 2012.

[ii] *Ibid.*, Courtesy of Guardian News & Media Ltd.

[iii] Eric John Ernest Hobsbawm (1917–2012) Marxist Historian

[iv] CND @campaigns to scrap nuclear weapons and create genuine security for future generations. Retrieved from http://www.cnduk.org/. Accessed 9 July 2016.

[v] Department for Business Energy and Industrial Strategy responsibilities for business, industrial strategy, science, innovation, energy, and climate change. Retrieved from https://www.gov.uk/government/organisations/department-for-business-energy-and-industrial-strategy.

[vi] Retrieved from http://www.nobelprize.org/nobel_prizes/economics/laureates/1984/stone-autobio.html.

[vii] John Maynard Keynes, 1st Baron Keynes; 1883–1946) British Economist, considered a founder of modern macroeconomics and most influential economist of the 20th century, his work informed the economic policies of governments. Retrieved from http://en.wikipedia.org/wiki/John_Maynard_Keynes.

[viii] Imperial College Industrial Sociology Unit developed in the Department of Mechanical Engineering.

[ix] See Joan Woodward entry.

[x] Imperial College Department of Social and Economic Studies developed in from.

[xi] See Sandra Dawson entry.

[xii] Lisl Klein (1928–2015) Organisational research expert had experience in factories, industry and HR management and later as social services advisor. After Imperial, she went on to work at the Tavistock Institute of Human Relations as Senior Social Scientist 1971–1989, and founded the Bayswater Institute in 1990, which "aims to improve the efficiency of organisations without damaging the quality of their members"lives; improve the quality of working life without damaging efficiency; where possible, to improve both simultaneously. Retrieved from http://bayswaterinst.org/. Accessed 9 July 2016.

[xiii] See Dot Grifitths entry.

xiv Trades Union Council (TUC) a national trade union centre, a federation of trade unions in England and Wales, representing the majority of trade unions, with wide lobbying and negotiating remit with government. Retrieved from http://en.wikipedia.org/wiki/Trades_Union_Congress http://www.tuc.org.uk/.

xv How it All Began; A Footnote to History Marion Kozak Socialist Register vol.31, 1995, pp. 264–285.

xvi *The Listener*, Historical Archive 1929–1991 Gale Historical Newspapers The Listener Historical Archive belongs to the Gale Historical Newspapers Collection

xvii Royal Holloway and Bedford New College Obituary. Retrieved from http://www.rhul.ac.uk/aboutus/newsandevents/news/newsarticles/tributetoprofessordorothywedderburn.aspx. Accessed 20160709.

xviii Reflections on the Merger of Royal Holloway and Bedford Colleges 1981–1985 Dorothy Wedderburn Higher Education Quarterly Volume 45 Issue 2 pp. 145–157, April 1991.

xix October 21st 1988 issue 814. Retrieved from https://issuu.com/rspall/docs/1988_0814_a. Accessed 20160709.

xx March 3rd 1989 issue 830. Retrieved from http://felixonline.co.uk/archive/IC_1989/1989_0830_A.pdf. Accessed 20160709.

xxi *The Independent*, Obituary Dorothy Wedderburn 20121005.

xxii Sir Walter Bodmer, Public Understanding of Science: The BA, the Royal Society and COPUS*. Notes and Records Royal Society Published online 14th July 2010.

*Royal Society's Committee on the Public Understanding of Science. Retrieved from http://rsnr.royalsocietypublishing.org/content/early/2010/06/16/rsnr.2010.0035.full.pdf+html.

Joan Woodward

Name and Post Nominals

Joan Woodward (1916–1971).

Image from the cut out style plaque installed in Woodward Buildings, the Hall of Residence named after her in 2015.

Specialism

Organizational Theory: Contingency theory, the relationship of technology to a business and its organization.

Career Roles

Roles in industry 1939–1946; Civil Service 1946–1948; Lecturer Department of Social Science Liverpool University 1948–1953; Director Department of Science and Industrial Research (DSIR) Unit in East Essex College of Technology 1953–1957; Department of Social and Administrative Studies as Special Tutor in Industrial Sociology Oxford University 1957–1962; Part time Lecturer in Industrial Sociology 1957–1962; Senior Lecturer Production Engineering Section 1962; Chair in Industrial Sociology 1970 Imperial College; National Board for Prices and Incomes Member, 1968–1971.

Awards

Travelling scholarship to the League of Nations Junior Summer School Geneva August 1933. The Joan Woodward Prize (awarded annually) organised by the Business School, funded by an endowment created in her name.

Student Hall of Residence, Imperial College named Woodward Buildings 2015.

Professional Affiliations

Lectures

In Her Name

Joan Woodward Memorial Lecture (Bi-annual) and the Joan Woodward Prize (awarded annually). Funded by an endowment created in her name at Imperial College.

Quote

> Joan epitomised the spirit of the College in its commitment to the application of ideas to practice.[i]

Born

19th June 1916–1971.

Family Background

Father: Joseph Henry Woodward; Mother: Lily Woodward (*née* King).

Marriage

Leslie Thomas Blakeman later CBE, (1904–1975) 1951. Labour relations expert. One step daughter.

Education

Schooling

Milham Ford School Oxford.

University

Undergraduate

University of Oxford.

Degree

BA.

Postgraduate

University of Durham.

Degree

MA.

Text Discussing Career and other Comments

Joan was the second woman professor at Imperial, appointed to the Chair in Industrial Sociology in 1970. The first woman academic to have a new Hall of Residence named after her, Woodward Buildings, Acton, opened in 2015. She joined the College in 1957 as a part time lecturer in Industrial Sociology, and was appointed to a Senior Lectureship in the Production Engineering Section in 1962. She was recognised as one of the world's leading organisation theorists.

Joan undertook major long-term studies into organisations, overturning some standard theories in relation to organisational structure and organisational performance. She also studied individual work roles such as dockers and saleswomen.

Joan's early life was in Oxford, she attended a girl's school, Milham Ford, in west Oxford. She was one of the six children chosen from schools in Oxfordshire to win a travelling scholarship to the League of Nations Junior Summer School in Geneva, August 1933. This was a time of fear of another war and ferment to prevent such a thing happening again. The competition was set up by the then editor of the *Oxford Mail* newspaper, Charles Fenby. He enlisted the assistance of Oxford Classics Professor Gilbert Murray,[ii] Chairman of the League of Nations Union, who set the essay titles: *Why I believe that the League of Nations can secure Disarmament, The League's Work for Humanity, War is inevitable: Answer this in the Light of the League's Work.* Murray judged the essays, selecting Joan and two other girls and three boys from the entrants.[iii] This was perhaps an early indication of Joan's interest in organisations.

Joan Woodward 2nd from left as a school girl, League of Nations Essay Prize Winner 1958.

Joan Woodward's career began in 1939 in industry, and then the Civil Service, before developing into a distinguished academic life. She was firstly a Lecturer in Liverpool University's Department of Social Science, then Director of the Department of Science and Industry (DSIR) Unit in Essex, before moving to Oxford's Department of Social and Administrative Studies as Special Tutor in Industrial Sociology. In 1957, she joined Imperial College as a part time Lecturer in Industrial Sociology, and was made fulltime Senior Lecturer in the Production Engineering Section in 1962.

The section that Joan joined at Imperial grew out of the Mechanical Engineering department's Production Engineering section, and the courses eventually made up the Department of Social and Economic Studies and Department of Management Science, merging into the Management School, and later the Imperial College Business School, although the sociological aspects were mostly subsumed.

Dot Griffiths who worked with Joan in the late 1960s and early 1970s, and cites her as a great mentor says:

> 'Joan's most famous work was a longitudinal study of 100 organisations begun before she joined Imperial College, completed at the College, and published as *Industrial Organisation: Theory and Practice*.[iv]

Her key and lasting insight was that technology is a key variable influencing organisational design and performance. Her sample of firms covered a range of manufacturing technologies and what she demonstrated was that the successful firms were the ones where there was congruence between the type of technology they were using and their organisational structures and processes. Different technologies generated different requirements and the successful firms recognised this. It is a tribute to Joan that this insight which was so controversial in its time is now part of the "taken for granted" world of the organisation theorist.

This work established Joan as one of the world's foremost organisation theorists. She was invited to join a group of the top seven theorists which called itself the Magnificent Seven. This international recognition was a huge achievement for a woman in the 1960s.[v]

Joan's expertise was sought by industrialists and government, and she was involved with many national committees. Her husband, Leslie Blakeman, worked in labour relations for the government and major companies, and was President of the Institute of Personnel Management 1965–1967. Their fields of work were therefore happily complementary.

They had no children, but Leslie had a daughter from a previous marriage.

Lisl Klein[vi], an expert in organisational theory and a social scientist who also worked with Joan liked her style of "go getting" which she applied to research, and to seeking people to work with, who would be assets to the department.

The Imperial College Memorial Lectures named after Joan have been given by eminent speakers in the field for 30 years. Therefore, Professor Dot Griffiths and colleagues who had known Joan, decided the time was right to honour Joan's memory in print, and so took on the task of editing a book of essays in her honour, a Festschrift.[vii] This book demonstrates her legacy through the continuation of her theories and their being put into practice up to present times.

Dot has the last word: "Joan played a significant part in the development of my own career as she appointed me to the College. She was a wonderful role model at a time when there were few senior women. She was also a warm and generous human being".

Professor Joan Woodward was a special woman who holds a place in the hearts of those that knew and worked with her.[viii]

Endnotes

[i] Professor Dot Griffiths. Imperial College Centenary Timeline 2007. Retrieved from http://www.imperial.ac.uk/centenary/memories/dotgriffiths.shtml.

[ii] George Gilbert Aimé Murray, OM (1866–1957). Retrieved from http://gilbert-murraytrust.weebly.com/about-gilbert-murray.html.

[iii] Private Communication. The Oxford Mail 11th August 1983. The Anthony Wood Diary Mission of peace in Shadow of War.

[iv] Oxford University Press 1965.

[v] *ibid.*

[vi] Lisl Klein, Interviewed November 2015.

[vii] Technology and Organization: Essays in Honour of Joan Woodward. Nelson Phillips, Graham Sewall, Dot Griffiths (eds.) *Research in the Sociology of Organizations*, Vol 29, 2010, Emerald Group Publishing Ltd.

[viii] *Op cit.*, i.

Dame Sandra June Noble Dawson

Name and Post Nominals

Professor Dame Sandra June Noble Dawson (1946–) DBE, PhD, FIPH FCGI, FACSSC, CIM, BA.[i]
First woman Master, Sydney Sussex College Cambridge.

Specialism

Business Management, Organisational Behaviour, Technology Transfer, Health Management, Health Policy.

Career Roles

Civil Service Research Officer Government Social Survey 1968; Research Officer Sociology Unit Department of Social and Economic Studies 1969, Lecturer 1977, Senior Lecturer 1984, Deputy Director Management School 1987–1994, Professor of Organisational Behaviour Imperial College 1990; Fellow of Jesus College 1995, KPMG Professor of Management Studies 1995–2003, Director Judge Institute of Management Studies 1995–2006, (now Cambridge Judge Business School) Master Sydney Sussex College 1999–2013; A Deputy Vice-Chancellor Cambridge University 2008–2012.

Non-executive Director Riverside Health Authority London 1990–1992; Chairman Riverside Mental Health Trust London 1992–1995; Member Research Strategy Board Offshore Safety Division Health and Safety Executive 1991–1995; Strategic Review Group PHLS 1994; Senior Salaries Review Body 1997–2013; Non-Executive Director Cambridge Econometrics 1996–2007; Fleming Claverhouse Investment Trust 1996–2003. Non-executive roles Barclays Plc 2003–2009; Alchemy Partners 2002–2004; (DRS) Data and Research Services plc 2012–; Oxfam 2006–2012; 2010–2013 FSA; Member, the Prime Minister's Council for Science and Technology;[ii] Member UK India Round Table.[iii]

Professional Affiliations

Fellow Institute Public Health (FIPH); Fellowship City and Guilds of London Institute (FCGI); Chartered Institute of Marketing (CIM).

Awards

Dame British Empire (DBE) "in recognition of her contribution to higher education and management research"; Honorary title of Companion of ABS (Association of Business Schools) International Women's Forum[iv] Global Hall of Fame induction, in recognition of achievements, including guiding the successful development of Cambridge Judge Business School; Hon DSc Keele University.

Quote

> Work and family life are two enormously important parts of my life and I have always been determined to do well at both.[v]

Born

4th June 1946.

Family Background

Marriage

1969 Henry R.C. Dawson; Children one son, two daughters.

Education

Schooling

Dr Challoners School Amersham.

University

Undergraduate

Keele University.

Degree

BA Hons History and Sociology.

Postgraduate

Cambridge University

Degree

MA

Text discussing Career and other Comments

Sandra Dawson has always maintained a down to earth view of her priorities in life — she has excelled at both professional and family life. Looking at a chronology of what she has achieved, it appears to be a very smooth path, but this disguises the everyday juggling of interests, organisational quirks and adaptations that have to be made by anyone.

She says in her interview to Cambridge University Press Office,[vi]

> …Successful management depends on all key parts working together; if any element changes (a new member of staff, a major overhaul of computing technology, for instance) then the structure and strategy must always be reviewed…

It would seem that Sandra has applied this system to her own life very successfully. It is rare that women in academic life leave their students and research projects without attention during maternity leave as academic needs require momentum to be maintained. Similarly, Sandra kept her work and home life going during what must have been a distressing time, when one of her small children had a serious accident and required extra care and support.

After graduating with a first class degree in sociology and history from Keel University, one of the new universities, 'red bricks' as they were known, Sandra Dawson began her career in the Civil Service as a Research Officer on a Government Social Survey in 1968. She was on the fast track, intending to spend her career in the Civil Service. However, it was not to her liking, the pace being too slow and so she changed track and went into academic life, joining Imperial College in 1969 as a Research Officer in the Sociology Unit in the Department of Social and Economic Studies. She was made a Lecturer in 1977, then Senior Lecturer in 1984, and, as the department evolved into the Management School, she became Professor of Organisational Behaviour in 1990. Between 1987 and 1994 she was the School's Deputy Director. Sandra moved to Cambridge University in 1995, being a Fellow of Jesus College, and until 2006 was Director of the Judge Institute of Management Studies, (now the Cambridge Judge Business School) where she is also KPMG Professor of Management. For 10 years from 1999, she was Master of Sydney Sussex College, the first woman to hold this position.

Simultaneously to her academic work, Sandra worked as a director or board member to trusts and organisations as noted above. They varied from health trusts, the Health and Safety Executive, financial organisations and governmental bodies. Working with these organisations contributed to her continued understanding of contemporary corporate life, essential to one who is teaching those from, or going out to, modern organisations. This experience "…allowed her to keep in touch with the complex issues of real management and put into practice what she preached.."[vii]

Besides contributions to academic journals, Sandra has published many books since the start of her career, including contributing with her colleague Dorothy Wedderburn,[viii] to their colleague Joan Woodward's book whilst they were all at Imperial College together: S. Dawson, and D. Wedderburn, (1980) "*Introduction: Joan Woodward*[ix] *and the development of organization theory*."

In Woodward, J.: Industrial organization: theory and practice. Oxford: Oxford University Press, 2nd edition, pp. xiii–xxvi.

Like other staff at Imperial, Sandra benefited enormously from the provision of the Day Nursery, in fact she was instrumental in setting it up, and in the previously quoted 1995 article says she realises she '…could never have done this [work] without having a very supportive husband and the benefits of Imperial's nursery.'

It was Sandra who with other career and family-oriented women at Imperial realised that it was essential to treat their lives as a whole, in order to successfully follow their chosen careers. It made sense to bring the family into the workplace, although at a remove, so that children were safe and cared for and work could be carried on uninterrupted without concern for their welfare. She has said that Imperial's "… very advanced nursery policy, enabled her, as a young mother, to carry on her research and teaching. Dropping her children off at the nursery before lectures, feeding them during breaks, and then taking them home when she finished work meant that she could stay on the career ladder and maintain a close eye on child-rearing…."[x]

Sandra has enormous energy and vitality — key features in those who are highly successful. Sandra is an example of a woman who has succeeded in all aspects of her life. Her children have grown up and she has developed her career in a way undreamt of by many women when she started out at school in the 1950s. She has been recognised internationally and nationally by the supreme award of Dame of the British Empire: something to aspire to.

Sandra takes part in events such as those organised by AWISE[xi] and Cambridge Business Women's Network, which are open to women who wish to hear about the careers of successful women and how they might emulate this success. There are discussions and break-out sessions for the attendees, often such events are held in schools and attended by sixth form girls who are making their career choices. Cambridge Business Women's Network website quotes her as saying that among her top tips for women in business, were to recognise and exploit opportunities and not to agonise over decision-making.[xii]

In an interview for McKinsey in 2008, she discusses how women can be effective in the board room, but these ideas are applicable in any career; she says: '… It is useful to have the checklist for women — speak up, be assertive, speak in short sentences, think carefully about your career moves and how you present yourself, etcetera — and checklists for companies to ensure equal opportunity. But let us have those as an enabling framework, not a rigid straitjacket. It would be such a pity if in tackling these issues to achieve diversity, we ended up with less diversity.[xiii]

Endnotes

i Image Sydney Sussex College website. Retrieved from http://www.sid.cam.ac.uk/aboutus/people/person.html?crsid=sjnd2.

ii The Council for Science and Technology (CST) is the UK *Prime Minister's* top-level independent advisory body on science and technology policy issues.

iii UK–India Round Table The UK–India Round Table was established by Foreign Secretary Robin Cook and Indian Foreign Minister Jaswant Singh in April 2000. They were looking for ways to regenerate and strengthen the bilateral relationship. Early meetings were chaired by Mr. KC. Pant and Lord Paul of Marylebone. The current Co-Chairmen are Lord Patten of Barnes and Mr. Nitin Desai. The Round Table consists of about 30 senior people, drawn equally from India and the UK, and include academics, businessmen, writers, politicians and journalists. All the UK members are invited personally by the Foreign Secretary. The two High Commissioners and a small number of officials from each side are *ex officio* members of the group.

iv The International Women's Forum advances leadership across careers, cultures and continents by connecting the world's most pre-eminent women of significant and diverse achievement. The IWF brings together an unprecedented global membership to exchange ideas, to learn and inspire, and to promote better leadership for a changing world. Retrieved from http://iwforum.org/.

v Extract from interview Cambridge University Press and Information Office 1995. Retrieved from http://www.admin.cam.ac.uk/univ/newsletter/.

vi *Ibid.*

vii Extract from interview Cambridge University Press and Information Office 1995.

viii Professor Dorothy Wedderburn [*entry/page no*].

ix Professor Joan Woodward [*entry/page no*].

x Extract from interview Cambridge University Press and Information Office 1995.

xi Association of Women in Science and Engineering.

xii Cambridge Business Women's Network. An Audience with Dame Sandra Dawson *Grapevine* Issue 19 December 2011. Retrieved from http://www.cbn.org.uk/.

xiii Seeing beyond the woman: An interview with a pioneering academic and board member, *Sandra Dawson addresses the changing role of women in business over the last 40 years.* Mary C. Meaney, September 2008. Sandra Dawson, a prominent expert on organizational change and leadership and a *The McKinsey Quarterly* 2008 http://businessbuilders.bz/seeing_beyond_the_woman.pdf.

Sue Birley

Name and Post Nominals

Sue Birley (1943–).
First Professor of Entrepreneurship Imperial College.

Director of Research; Professor of Entrepreneurship, Management School (Business School).

Research topics: Small businesses; Entrepreneurship; Technology Foresight.

Sue worked in academic institutions in the UK and abroad, notably in the UK, at Cranfield School of Management, where she continued her research into gender difference in entrepreneurship with the paper entitled: *Female Entrepreneurs — Are they Really any Different?*[i] and Imperial College Management School, where in 1998 she was leading the entrepreneurship course.

In 2001 with Sue as Head, Imperial's Entrepreneurship Centre was launched to develop the entrepreneurial skills of Imperial's students. Sue said:

> All students now have the chance to learn about entrepreneurship and to try out their new business ideas. We believe that our Centre will widen the opportunities open to them, whatever they choose to do in the future.

> Our aim is to enhance the image of Imperial as the university for aspiring entrepreneurs and inventors.

The Centre also worked with academics on spinout companies.

Endnote

[i] The conclusion is that the major difference is in the market — entry choices made, *Journal of Small Business Management School*, Working Paper 5/87. Copyright: Birley, 1988.

Dorothy Griffiths

Name and Post Nominals

Dot Griffiths
OBE, MSc, BSc.

Specialism

Human Resource Management

Career Roles

Lecturer Industrial Sociology; Professor; Dean of Business School; Head of Business School Programmes Acting Head Business School; Professor of Human Resource Management and Head of Programmes; Athena Swan Champion, Provost's Envoy for Gender Equality Imperial College; Chair of the Central and North West London NHS Foundation Trust (CNWL); Founding Editor *Feminist Review*; Chair of Trustees of the Feminist Review Trust.

Career Awards

Fellow City and Guilds of London Institute (FCGI) 2004; Fellow RSA 2010; Member Higher Education Academy 2003.[i]

Awards

OBE for Services to Higher Education 2010.

Quote

> I've always believed in leading from the front, I like to get things done; if you're a leader, lead.[ii]

Born

London, 1947.

Family Background

Father: Henry Griffiths, Engineering Patternmaker; Mother: Jacky Griffiths (née Broderick) Typist.

Marriage

None.

Education

Schooling

Brondesbury and Kilburn High School for Girls.

University

Undergraduate

University of London.

Degree

BSc Sociology.

Postgraduate

Bath University Sociology of Science and Technology.

Degree

MSc.

Text Discussing Career and Other Comments

Dot Joined Imperial College in 1969, at a very exciting time for the development of sociology, both externally and in Imperial College. The Industrial Sociology Unit was essentially run by women of outstanding calibre[iii] who had made extensive impacts on academic sociology, developing it as an academic discipline in the fields of labour relations and practical applications to the work place. This was in line with the mission (and tradition) of Imperial College and its applications of technologies and systems to industry through its staff and students.

Dot made her mark both inside and outside Imperial College, and was at the core of radical thinking about the matter of social responsibility in scientific work and women in society.

In the British Society for Social Responsibility in Science,[iv] Dot worked with its founders including Maurice Wilkins (Nobel Prize winner for discovery of DNA) the founding President from 1969 until 1991, when the organisation closed down. Other members of the Imperial community included a number of staff in the Physics Department including Tom Kibble, (copropounder of the existence of the Higgs Boson) and student members such as Student Union President Judith Walker and later, Mary Attenborough.

Dot was a founding editor of the journal *Feminist Review*[v] in 1979 (the100th issue of the journal was the subject of analysis by its current collective editors in 2013)[vi] and is Chair of Trustees of the Feminist Review Trust.[vii]

One of the aspects of working life Dot has consistently been concerned with is that of women and their roles in work, particularly in management.

Dot set up the UK's first ever Women in Science Group.[viii] "I was a founding member and the only woman from Imperial in the group. I felt very strongly that I needed to bring this into the College if I could." This group pioneered discussions of the social dynamics of science from a feminist perspective and led Dot to a lifelong concern about understanding and challenging the over-representation of men in STEM subjects, particularly at Imperial.

She has been at the forefront of developing support for women at Imperial College through both national and local initiatives and championed the cause through Imperial's Academic Opportunities Committee which she chaired. Through this she established the Faculty Women Ambassador's Scheme — a senior woman in each faculty to champion women in her faculty. She led the College level interaction with the Athena SWAN project and charters. The College was a founding signatory to the charter. Amongst her other initiatives were the Julia Higgins Medal and Prizes, the Women@Imperial Week and a system of sessions designed to assist both women and men academics in thinking about their applications for interviews and promotions.

Dame Julia Stretton Higgins[ix] says of Dot:

"She became the main driver for getting these initiatives implemented. There were plenty of good people with lots of ideas but you need a real leader to get things done, and Dot was just that."[x]

Dot attributes some of her success to being 'naturally bossy', but it is more than that — it is the combination of her natural intelligence and common sense which she applies to everything she does that makes things happen. The prestigious Athena Swan[xi] programme has also been keenly promoted by Dot to great acclaim in many departments.

Dot's work evolved with the Industrial Sociology Unit into the Management School which then became the Business School. She was made professor in 2002 and was Deputy Dean and Head of Programmes in the Business School, which meant that she was responsible for the design and delivery of the school's programmes, for teaching quality and for school accreditations. She more recently took on the role of Dean of the business school, shortly before retiring from that side of her academic life.

Her influence on the success of the business school is recognised by her successor as Dean, Professor G. 'Anand' Anandalingam "Dot really helped the Business School reach new heights in terms of student enrolment, women now make up almost half of the students enrolled on the full-time MBA programme at the Business School."[xii]

As a senior academic, Dot was a member of College strategic committees, which have oversight of the College mission, ensuring it adhered to it in all aspects and reviewed and developed as necessary.

The title of Professor of Human Resource Management is retained by Dot as she transfers part time to a role in College Management. Dot remains the Provost's Envoy for Gender Equality at Imperial College, a role she takes very seriously and for which she is ideally suited, with her broad working experience and background in gender attitudes.

Dot had many external roles during her full-time career at Imperial, and is indeed still in demand as an international consultant in Human Resource Management, having worked, for example for the BBC, BP, Institute of Directors, National Power, Securicor, the Institute of Development Studies and Unilever. She has interests in talent acquisition and talent development and has also worked on quality measures for HR performance and as an Executive Coach.

On retiring from full-time work at Imperial, Dot became Chair of the Central and North West London NHS Foundation Trust (CNWL) on 1st January 2014. She had been a Non-Executive Director at CNWL since 2000 and the Trust's Senior Independent Director since 2007. This means that as she enters the role she has a very good working knowledge of the Trust and of its members and governors, so has a robust position from which to develop her role.[xiii] She will no doubt apply her vast knowledge of people and organisations and how they tick, her sense of humour, approachability and gently forceful persuasive energies to the success of this organisation as she always has done, and continues to do, at Imperial.

Endnotes

[i] The Higher Education Academy champions excellent learning and teaching in higher education. Retrieved from http://www.heacademy.ac.uk/home

[ii] A Fond Farewell to Professor Dorothy Griffiths Maxine Myers Communications and Public Affairs Imperial College News 16th December 2013. Retrieved from http://www3.imperial.ac.uk/newsandeventspggrp/imperialcollege/news-summary/news_16-12-2013-16-4-.

iii Dorothy Wedderburn, Sandra Dawson, Joan Woodward, whose biographies appear in this volume.

iv British Society for Social Responsibility in Science For the Archive and a description of its work. Retrieved from http://www.bssrs.org/

v *Feminist Review* is a peer reviewed, interdisciplinary journal setting new agendas for feminism. Produced by a Collective, the journal is committed to exploring gender in its relationship to other axes of power including race, class and sexuality. Retrieved from http://www.springer.com/social+sciences/journal/41305

vi A review of Feminist Review's 100th issue: Celebrating 100 issues of collective practice By Hamilton, C (Hamilton, Carrie) [1] ; Hemmings, C (Hemmings, Clare) [2]; Al-Ali, N (Al-Ali, Nadje) [3] *EUROPEAN JOURNAL OF WOMENS STUDIES* 20(1) pp. 93–99, 2013.

vii Retrieved from http://www.feminist-review-trust.com/ Supports initiatives that transform the lives of women.

viii Retrieved from http://www3.imperial.ac.uk/newsandeventspggrp/imperial-college/newssummary/news_16-12-2013-16-4-51.

ix See Julia Higgins entry.

x *Ibid.*, ii.

xi See Athena Swan references.

xii *Ibid.*, ii.

xiii Central and North West London NHS Foundation Trust is an NHS Foundation Trust in England. It was created in 2002 by a merger between Brent, Kensington & Chelsea and Westminster Mental Health NHS Trust, Harrow and Hillingdon Healthcare Trust, and the substance misuse service component of Hounslow and Spelthorne Community and Mental Health NHS Trust. It subsequently won additional contracts, including Milton Keynes Community Health Services from April 2013. [1] It has substantial contracts for prison health services. Retrieved from http://en.wikipedia.org/wiki/Central_and_North_West_London_NHS_Foundation_Trust.

Principals and Vice Chancellors

Frances Mary Gore Micklethwait

Name and Post Nominals

Frances Mary Gore Micklethwait (1867–1950)
MBE, ARCS 1901, FCS, AIC, FIC.
First woman Mature Student in Chemistry, Royal College of Science.
The Frances Micklethwait Centenary Memorial Prize, 2007.

Specialism

Organic Chemistry.

Career Roles

Private Assistant to Professor G.T. Morgan, Chemistry Department 1901–1914; Research Chemist Boots Pure Drug Company; Inspector of Munitions; Principal Swanley Horticultural College for Women; edited Thorpe's Dictionary of Applied Chemistry with her former colleague Martha Annie Whiteley.

Career Awards

Beit Memorial Fellowship for Medical Research 1914; MBE for WWI work 1919; Fellow Chemical Society 1920; Associate 1904, Fellow 1907 Institute of Chemistry.

Awards in Her Name

The Frances Micklethwait Centenary Memorial Prize was created in her name at Imperial College. This prize is awarded to a UG student who shows outstanding performance over the course of the degree and who produces an excellent research project.

Quote

> Her research remains to remind us of a fine chemist, F.H. Burstall.[i]

Born

7th March 1868.

Died

25th March 1950.

Education

Privately

Swanley Horticultural College

Royal College of Science, Imperial College 1898–1901 ARCS.

Career and Other Comments

Coming to the Royal College of Science at a later age than most, being 30, so very much a mature student, Frances worked closely with a professor of chemistry, G.T. Morgan, between 1902 and 1914. The department was small and the men and women, staff and students, must all have been in close communication. Subsequently she worked with M.O. Forster and J.C. Cain in the lab, again with a good research output. Martha Annie Whiteley was also one of her close colleagues. Frances was a great collaborative research chemist. From the start of WWI in 1914, she worked in chemical munitions research becoming Inspector of Munitions in 1916. Postwar, her employment was in the research department of Boots Pharmacists and then at Swanley Horticultural College for Women, becoming Principal in 1922. However, she remained loyal to the department in continuing to contribute to the text book instigated by Sir Jocelyn Thorpe: *Thorpe's Dictionary of Applied Chemistry*, a work that ran into several editions. Martha Annie Whiteley took on the editorship after Thorpe's death, having been involved with its preparation from the first.

Endnote

[i] Obituary Journal of the Chemical Society, 1952, 2946–2952. Birth date corrected to 1867 by Tracey Logan personal communication.

Kate Barratt

Name and Post Nominals

Kate Barratt (1884–1977)
CBE, DSc.
Principal Swanley Horticultural College

Specialism

Botany.

Career Roles

Demonstrator; Lecturer Swanley Horticultural College; Demonstrator; Lecturer Imperial College; Organiser for the Women's Land Service Corps; Principal Swanley Horticultural College.

Career Awards

CBE; Fellow Wye College.

Awards

Swanley Silver Salver first in exams 1904.

Family Background

Marriage

1950, to Richard John Tabor (1875–1958) botanist, following their long term prior relationship.

Lecturer at Swanley Horticultural College (1899–1910).

Education

University

Undergraduate

Swanley Horticultural College Diploma 1904; Imperial College 1910–1913.

Degree

BSc.

Undergraduate Awards/Scholarships

National Scholarship to attend Royal College of Science Botany Department Imperial College.

Postgraduate

Imperial College 1912–1913.

Degree

MSc 1913; DSc 1920.

Text Discussing Career and Other Comments

Kate's botanical career began with her training at Swanley Horticultural College, and she retained an interest in Swanley throughout her life, becoming its last Principal.

An outstanding student, she came first in exams for her Diploma, awarded after the two year course, and also won therefore, the silver salver. She came second in the Royal Horticultural Society exams. Following the award of her Diploma in 1904 Kate became a botany demonstrator there, working with her very much in the future husband to be, Richard Tabor — they did not marry until 1950, but followed remarkably similar career paths at Swanley and Imperial.

Wishing to continue her studies to degree level, in 1906, Kate gained a National Scholarship to the Botany Department of the Royal College of Science (RCS) Imperial College where she came under the tutelage of Professor John Bretland Farmer, graduating BSc. She then returned to Swanley, just as Richard Tabor moved to Imperial.

Kate took up the lectureship in botany at Swanley, returning in 1912 to Imperial to complete her MSc. She then worked with Farmer as Demonstrator until 1920, when she was appointed Lecturer, and gained her DSc. Concurrently, Kate became a member of Swanley's Governing Body and also edited its *Horticultural College Magazine*. Kate and Richard worked on papers together sharing an interest in the cytology of plant stems and a fungal disease of the beech tree.[i]

Kate undertook research in Botany at Imperial between 1912 and 1913, joining the fourth year course one day per week. She was appointed Demonstrator in October 1913, which meant that she would assist with teaching and also support students in their laboratory work, at a salary of £150.00 per year. This remained the same until in 1919 she was paid £200.00, rising to £250.00 in 1920.

Her period of employment covered that of the WWI, and her expertise was required in active service as organiser for the Women's Land Army, or Service Corps. This Corps had been formed in 1916 and was funded by the Government. There were not enough women working in agriculture to replace the men who had been called up to fight, but those women who joined the Corps made a huge contribution to the continuation of agricultural output during the war. It was essential that women such as Kate, with agricultural knowledge and organisational skills, able to motivate and create teams to successfully, that is correctly, harvest crops, were recruited. Kate was asked by the head of the Corps to assist.

During the last part of 1918, she was granted leave of absence by her Professor, John Bretland Farmer and the College Secretary, Mr Alexander Gow offered to make up the shortfall in her salary for this war work.

Kate had discussed the subject of secondment with Professor Farmer before writing officially requesting leave of absence on the 26th January 1918[ii]:

> 'As we have discussed verbally already, I have been invited by Mrs Roland Wilkins the Chairman of the Executive Committee pf the Women's National Land Service Corps to undertake work in connection with the organisation of camps for temporary work by women, of the educated class, who will be flax pulling and first picking during the coming season. ...Mrs Wilkins particularly desires my help because of the experience I gained organising the camp for raspberry pickers last year in Scotland and also because of my familiarity with agricultural conditions. I shall be glad therefore if you will grant me leave of absence for the second half of this session. It is of course clearly understood that I shall be ready to recommence my duties in October next.
>
> Sincerely, Kate Barratt'
>
> Professor Farmer wrote to Mr Gow:
>
> 'As this work is of national importance, dealing as it does with the organisation of women to flax pulling, first gathering etc, I would recommend very strongly that the difference (£2.00 per week) between the salary here and the remuneration she gets from the Land Service Corps be made good out of my vote. I should not have willing consented to her leaving at all had I not been satisfied that it was in the national interest that she — with her rather unusual gifts of organisation — should temporarily leave us.'

Flax was an important crop as it was used in linen cloth making and also for canvas and webbing, materials essential to the functioning of the military campaigns e.g. in tent making, therefore it was necessary to ensure its harvesting and in the correct way to make it usable. The grains are edible and a source of protein, so it is a valuable dual purpose crop.

Kate undertook this work and then returned to Imperial and briefly became a lecturer in the department, resigning this position to take up the post of Principal of Swanley Horticultural College. In her letter she wrote:

"I would like to say how very much I regret leaving the College having thoroughly enjoyed my work under Professor Farmer."[iii]

Kate remained at Swanley for 22 years, the College was lost as an independent entity following its destruction by bombing in 1944, during WWII. Kate managed the evacuation and the subsequent necessary merger with the South Eastern Agricultural College at Wye College in 1946, and retired in 1947. She was made a Fellow of Wye College in 1949.

The career relationship between Kate and Richard demonstrated how they could manage their work and long term friendship culminating in marriage, each alongside their own careers: he became an Assistant Professor at Imperial, whilst she became Principal of Swanley and was awarded the CBE for her war work.

Endnotes

[i] Opitz D Article: '*A Triumph of Brains over Brute*': *Women and Science at the Horticultural College, Swanley, 1890–1910* Isis (2013).
[ii] Imperial College Archives.
[iii] *Ibid.*, ii.

Sources: General Information Wye: *The Journal of Wye College Agricola Club* **18**(4) 2014 pp. 18–19.

D. Opitz *Swanley Magazine*, May 1946, **6**(97), 7–8.

Department of Science and Art of the Committee of Council on Education, General Register of the horticulture College, Swanley, Kent, SWAN00007, HHC; and [Botany Department], Students Register, 1900–1922/23, KB/19/1, Imperial College Archives.

Jenny Higham

Name and Post Nominals

Jenny Higham (1961–)
MRCO, FRCOG, MBBS, MD.
Principal St George's University of London 2015–.
First woman Principal St George's University of London.

Education

MBBS 1985, MD 1993 University College London
MRCOG 1992, FRCOG 2005.

Career Roles

Senior Registrar, St Mary's Hospital 1995–1996; Consultant Surgeon Gynaecologist, NHS and Private Practice 1996–; Head, Undergraduate Medicine, Imperial College, 2006–2009. Senior Vice Dean, Lee Kong Chian School of Medicine, Singapore, 2013–.

Imperial College 1995–2015

Vice Dean for Institutional Affairs and Director of Education in the Faculty of Medicine at Imperial College.

Deputy Principal Faculty of Medicine 2015.

'Mentor of the Year' award at the National Women of the Future Awards 2011.

President & Rector's Medal and Award for Outstanding Contribution to Teaching Excellence 2013.

Imperial College Medal for outstanding leadership 2014.

Lee Kong Chian School of Medicine in Singapore 2009

London Singapore Board 2009.

Senior Vice Dean of the Lee Kong Chian School of Medicine in Singapore 2013.

Spearheaded negotiations for the creation of the Lee Kong Chian School of Medicine in Singapore, the newest medical school in Singapore which awards a joint Imperial College and Nanyang Technological University (NTU) medical degree.

Lee Kong Chian School of Medicine opened 2013.

Nanyang Education University Gold Award and Fellow of the Teaching Excellence Academy 2015.

Educator of the Year at Nanyang Technological University's (NTU).

Joint Imperial College and Nanyang Technological University medical degree.

One of the first of two recipients of the "The crème de la crème of NTU's teaching talents" most prestigious award for teaching excellence, established to recognise "superlative excellence in teaching" 2015.[i]

Research

Reproductive medicine — a new method of menstrual blood loss assessment (Pictorial Blood Loss Assessment Chart) which has changed research practice. Original work on abnormal menstrual bleeding developed her reputation as a widely recognised expert, contributing to national and international definitions.

Simulation in education involving collaborations with engineers and veterinary practitioners.

Clinical expertise in the management of abnormal uterine bleeding, outpatient hysteroscopy and reproductive surgery.

Married

Edward George Naylor 1992, Children: one son two daughters.

Quote

From my personal perspective (Imperial) College has been supportive. I don't think that is a platitude, I think it's a genuine desire to see people from a range of backgrounds — culturally, ethnically, gender-wise, develop and achieve their potential.[ii]

Endnotes

[i] Retrieved from http://www3.imperial.ac.uk/newsandeventspggrp/imperialcollege/newssummary/news_13-3-2015-16-35-57. Accessed 8 July 2016.

[ii] Imperial College Reporter, 5th December 2011.

Debra Humphris

Name and Post Nominals

Debra Humphris (1959–)
PhD.

Vice Chancellor University of Brighton 2015–.

First woman Vice Chancellor University of Brighton.

First Athena Lecture University of Brighton 2016.

Professor Debra Humphris Pipeline Engineering — Women in STEM.[i]

Education

Registered Nurse; PhD Implementation of policy into clinical practice 1999.

Career Roles

Registered Nurse

Senior Research Fellow, South Thames Regional Health Authority and Senior Research Fellow, Health Care Evaluation Unit, Department of Public Health Sciences, St George's Hospital Medical School.

University of Southampton

The New Generation Project Lead: Development of interprofessional learning in health and social care 2000.

Director, Health Care Innovation Unit 2003.

Focus on education innovation in health and social care.

Professor of Health Care Development.

Pro Vice Chancellor (Education and Student Experience), 2008.

Chairman, Employment and Skills Board for the Solent Local Enterprise Partnership

Imperial College 2012–2015

Pro-Rector Education 2012.

Responsible for the creation, leadership and delivery of the university's education strategic plan, which emphasised enhancing the student experience and the importance of graduate employability.

Vice Provost Education 2013–2015.

Overall responsibility for educational strategy and the quality of the educational provision across the university.

University of Brighton Athena lecture will be on the pipeline that enables women to take up STEM careers (Science, Technology, Engineering and Maths).

As an openly LGBT senior leader in higher education, Debra is passionate about equality and diversity.[ii]

Endnotes

[i] The first University of Brighton Athena lecture will be on the pipeline that enables women to take up STEM careers (Science, Technology, Engineering and Maths) 7th March 2016. Retrieved from https://www.brighton.ac.uk/about-us/news-and-events/news/2016/02-22-university-of-brighton-athena-lecture.aspx. Accessed 17 August 2016.

[ii] Brighton University. Retrieved from https://www.brighton.ac.uk/about-us/governance-and-information/management-board/index.aspx. Accessed 17 August 2016.

Julia Clare Buckingham

Name and Post Nominals

Julia Clare Buckingham (1950–)
BSc, PhD, DSc, FRSB, FRSA, HonFBPhs, FCGI.
Vice Chancellor and President Brunel University, London, 2012–.
First woman Vice-Chancellor Brunel University, London.

Award in Her Name

Julia Buckingham Prize Imperial College School of Public Health

> Imperial's Blyth Music and Arts Centre enhances life at College, enabling a happy combination of science and music.[i]

Education

BSc Zoology University of Sheffield 1968–1971; PhD Pharmacology 1972–1974, Dsc, 1987: (Royal Free Hospital School of Medicine University of London); London Business School (Accelerated Development Programme) 2008–2009.

Career Roles

Research Assistant Department of Pharmacology, Glaxo Laboratories Ltd 1971–1972.

Royal Free Hospital School of Medicine

Postgraduate Research Student Department of Pharmacology 1972–1974, Research Fellow, Honorary Lecturer Neuroendocrine Unit 1974–1980, Senior Lecturer Academic Department of Pharmacology 1980–1987 Royal Free Hospital School of Medicine

Charing Cross and Westminster School of Medicine and Faculty of Medicine Imperial College London

Professor of Pharmacology/Head Department of Pharmacology 1988–1997: Assistant Dean (Preclinical) 1992–1997 Charing Cross and Westminster Medical School; 1997–2003: Professor of Pharmacology, Head, Department of Neuroendocrinology, Deputy Head, Division of Neuroscience and Psychological Medicine 1997–2003 Faculty of Medicine Imperial College London; 2000–2003: College Dean for Non-Clinical Medicine 2000–2003, Head, Division of Neuroscience and Mental Health and Professor of Pharmacology 2003–2007.

Pro-Rector (Education) and Head of the College Centre for Integrative Mammalian Physiology and Pharmacology, 2007–2010.

Pro-Rector (Education & Academic Affairs) Head of the College Centre for Integrative Mammalian Physiology and Pharmacology, 2010–2012.

Responsible for leadership of the development and implementation of the College's overall educational strategy, the quality of educational provision and the direction of the academic support divisions which underpin the College's educational mission.

Chairman Centre for Integrative Mammalian Physiology and Pharmacology.

Research: Cortisol, the body's major stress hormone plays an essential role in the maintenance of normal bodily functions and but also contributes to the pathogenesis of disease when its pattern of secretion or activity is disturbed. My research has advanced understanding of the molecular mechanisms regulating cortisol secretion and its diverse actions in the body.

Married

Simon James Smith, 1974.

Endnote

[i] I play the piano and it provides marvellous opportunities for students and staff to develop their musical and artistic skills.

Julia Elizabeth King

Name and Post Nominals

Julia Elizabeth King

Baroness Brown of Cambridge.

DBE, CBE, FREng, FIMMM, FRAeS, FIMarEST, FInstP, FCGI, FEI.

Specialism

Engineering, structural materials fatigue and fracture, aerospace and marine propulsion technology, Carbon Emissions and Environmental impact, Climate Change, Women in STEM.

Career Roles

Rolls-Royce Research Fellow, Girton College Cambridge University 1978–1980, Lecturer University of Nottingham 1980–1987; British Gas Fellowship of Engineering, Senior Research Fellow University of Cambridge 1987–1992; Teaching Fellow, Churchill College University of Cambridge 1987–1994, 2002–, Lecturer 1992–1994; Assistant Director Technology Centre for Ni-Base Superalloys 1993–1994; Head of Materials Rolls–Royce Aerospace Group 1994–1996; Director Advanced Engineering Rolls-Royce Industrial Power Group 1997–1998; Managing

Director, Fan Systems Rolls-Royce plc 1998–2000; Director Engineering and Technology Marine Rolls-Royce plc 2000–2002; Chief Executive, Institute of Physics 2002–2004; Principal, Faculty of Engineering Imperial College London 2004–2006; Vice Chancellor and Chief Executive, Aston University 2006–2016; Chair Henry Royce Institute for Advanced Materials University of Manchester 2015–.

Member

Board of Universities UK Chair Innovation and Growth Policy Network; Councils Engineering and Physical Sciences Research Councils; Member Board of the Directors of the National Centre for Universities and Business.

Government Advisor

Chair of the Defence Scientific Advisory Council Ministry of Defence; Cabinet Office National Security Forum; Science and Technology Honours Committee; Technology Strategy Board; King Review of vehicle and fuel technologies to reduce carbon emissions over 25 years, Reported 2008; UK's Low Carbon Business Ambassador; Airports Commission; Inaugural member on the Governing Board of the European Institute of Innovation and Technology, World Economic Forum Automotive Council; Deputy Chair Climate Change Committee.

Career Roles

Leader Royal Academy of Engineering Working Party on Educating Engineers for the 21st Century: Reported 2007. Browne Review of Higher Education Funding and Student Finance : Reported 2010; Stern Review of REF reporting Autumn 2016.

Career Awards

FREng 1997, FIMMM, FRAeS, FIMarEST, FinstP, FCGI, FEI; CBE Services to Materials Engineering 1999; DBE Services to Higher Education and Technology 2012; Baroness Brown of Cambridge 2015.

Medals

The Grunfeld Medal 1992, Bengough Medal 1991 Institute of Mining, Materials and Metals; Kelvin Medal 2002 Institution Civil Engineers, John Collier Medal 2010 cosponsored by Institution Chemical Engineers (IChemE), the Royal Society and the Royal Academy of Engineering (for exceptional commitment and contribution to nuclear industry and chemical engineering), Lunar Society Medal 2011 (for contributing significantly to the vigour of the West Midlands Community) and the International Congress on Fracture ICF Constance Tipper Silver Medal 2013 (for fatigue and fracture research).

Prizes

The Erna Hamburger Prize Ėcole Polytechnique Fėdėral de Lausanne 2013; President's Prize Engineering Professors' Council 2012.

Awards

Liveryman Worshipful Company of Goldsmiths 1998, Freeman City of London 1998; Honorary Graduate, Queen Mary University of London, University of Manchester, University of Exeter, University of Strathclyde, Aston University; Honorary Fellow, Murray Edwards College Cambridge, Cardiff University, The Society for the Environment, The British Science Association, The Polymer Processing Academy India, IET, IOMMM; Visiting Professor Graduate School Chinese Academy of Sciences.

Professional Affiliations

Royal Academy of Engineering, Royal Aeronautical Society; Institute of Physics; Institute of Mining Metals and Materials, Women's Engineering Society Role Model.

Lectures

sixth John Collier Memorial Lecture; Higginson Lecture Durham University 2006; Campbell Lecture University Southampton 2008.

Quote

> If we as a society aim for equality in science, technology, engineering and maths (STEM) careers and stop thinking in terms of a gender divide, we can start to even up the balance to the benefit of individuals and businesses.[i]

Born

11th July 1954 Kensington London.

Family Background

Father: Derrick, Medical Instrument Salesman; Mother: Jane, Teacher; Brother Tim, Engineer.

Marriage

1984 Dr Colin Brown, Engineering Director Institution of Mechanical Engineers.

Education

Schooling Hallfield School Hallfield Estate Bayswater

Godolphin & Latymer School Hammersmith London

University

Undergraduate

New Hall later renamed Murray Edwards College University of Cambridge.

Degree

BSc Natural Sciences specialising in materials science 1975.

Undergraduate Awards/scholarships

New Hall Posener Scholarship 1973.

Postgraduate

Murray Edwards College University of Cambridge.

Degree

MA; PhD Fracture Mechanics 1978.

Text Discussing Career and Other Comments

Her family background has given her ambition, a sense of perfectionism and a capacity for design and colour — the very attributes helpful to an engineer.

Her parents separated when she and her older brother were very young, she then lived with their mother in Hammersmith, and her brother with their father in Bayswater, so not far apart in west London. Julia's mother, Jane, was engaged in social issues, and passed a lot of the ideas of equality and social mobility onto Julia. This is reflected in Julia's work at Aston University which she says is a key higher education institution for widening participation from low socio-economic groups and ethnic minorities, a record she is very proud of.[ii]

Julia's parents were both highly intelligent but perhaps thwarted in ambition. Her father had not had a formal education, but her mother had attended Oxford and aspired to be an actress — she did become a theatre then ballet critic for the *Daily Worker* which resulted in theatre visits for Julia, but her main job was as a teacher. She taught at the local Hammersmith FE College and Julia spent some time there after school hours taking pot tery courses. They had little money to spare, so Julia made a lot of her own clothes and therefore a sewing machine was important. This meant it had to be maintained, and it fell to Julia to repair it, as otherwise it was long walk with a heavy object to Shepherd's Bush and the repair shop! This was her first experiment in engineering, though making pots would have given her practical knowledge of how materials behaved given different consistencies and handling techniques. By dint of the two women living alone,

Julia also had to work out how other items were to be repaired, so her career path could be said to have been set from an early age!

Julia also spent a lot of time close to Imperial in fact, at the Victoria and Albert (V&A) Museum and the Science Museum. At the latter, she admired the great model trains which could be made to move by turning handles, even though they were in glass cases. Another hint at her coming career perhaps, and then at the V&A the intricacies of jewellery and glass ware were fascinating to her.

Despite Julia and her brother Tim living apart, she was very competitive with him academically. He too was a gifted student and became an engineer with a design interest, and taught at universities but took early retirement running his own consultancy.[iii] Filling in her university application forms Julia wanted to put 'professor 'as her ultimate career choice.[iv]

At secondary school Julia enjoyed chemistry — which at that period was taught in a very hands on way, which made it exciting and also enjoyed physics. Her school had taken on the experimental Nuffield system of physics and chemistry teaching, which involved experimentation and then looking at results to see what the student makes of them. Much akin to a system in engineering called Conceive-Design-Implement-Operate (CDIO) Julia suggests.[v]

She found science at school fun, and she was good at it and has continued to be so! However, when she got to Cambridge University to take natural Sciences, she found the way physics as taught tedious, where she had previously been excited by it. Visions of being at CERN and drinking champagne to celebrate a new atomic particle discovery which she had found so attractive dissipated. At school she was inspired by other scientific developments such as supersonic flight — Concorde being created in the UK and France, space travel, and development of nuclear power — a clean fuel. By university, Julia and her fellow students were becoming concerned about the detrimental effects on the environment of science and technology. They read reports such as that of The Club of Rome, which suggested that natural resources would run out in the 21st century if action was not taken. The impact of work like this surely influenced Julia in her commitment to working with government on environmental matters in her subsequent career.

Besides studying physics, Julia also enjoyably studied chemistry and crystallography, and in her final year, metallurgy too. She knew she wanted a career as an academic, and to take a PhD as the next step.

As an undergraduate, Julia attended New Hall, which she was glad of, as it was more modern than the traditional women's Colleges. She recognised that her interests lay in structural engineering and the failure of structures e.g. fracture in metals and other materials. Crystallography and metallurgy were large components of this area of work. Her PhD work was sponsored by Harwell's, she worked on embrittlement behaviour of nuclear reactor pressure vessel steels. Her supervisor was John Knott (1928–) Materials Scientist, who was a great influence and is a life-long colleague. Both he and Julia were awarded prizes by the International Congress on Fatigue at a ceremony in 2013. He was awarded an Outstanding Life Achievement medal and Julia the ICF's inaugural Tipper Prize.[vi] This is named after another woman described in this book, Constance Fligg Tipper, who was a pioneer in metal fatigue. That a prize remains named for her demonstrates C.F. Tipper's great contribution to the science, and that Julia won it, her contribution too.

Following completion of her PhD, Julia took up a Post Doctoral Fellowship funded by Rolls-Royce at Girton College, Cambridge, and then became a Lecturer in Materials Science at Nottingham University, 1980–1987 and gained the first of the Royal Academy of Engineering's senior research fellows, funded by British Gas. She was then asked to return to Cambridge as a Lecturer in materials, and retained her Fellowship. Julia gained further kudos by successfully bidding for the department to become the Rolls-Royce university technology centre for nickel-based alloy development.

She moved over to Rolls-Royce, managing their materials department and this enabled her to see things from the customer point of view, rather than as an engineer, rightly proud of their engines. The superb Rolls-Royce engines had to power huge aircraft. American Airlines was one of their customers. Julia was seconded to them and saw the maintenance and spare parts supply system first hand. She fed back information to Rolls-Royce prompting them to improve customer service. After two years she was promoted at Rolls-Royce to Director of Advanced Engineering for the

Industrial Power Group, later becoming Managing Director of the Fan Systems Business, and then Engineering Director for the Marine Business.

Of working at Rolls-Royce she says: "I was one of the two or three most senior women there for my entire time with the company, and there was a recognition that we were only tapping into half of the talent pool."[vii]

Julia demonstrates a great interest in the careers of young people, both male and female, in engineering. In 2014 she wrote in the Guardian about the role of apprenticeships, suggesting that the Higher Apprenticeships would enable young people to experience higher education without the expense of the usual degree course route. In taking a Higher Apprenticeship, they had an income as well as good training. Sponsored by a company which had developed a suitable five year course in concert with a University, in their first two years they gain a Higher Apprenticeship, equivalent of a first year pass at degree level, besides proving their practical capabilities and commitment to their subject and employment, and can then continue to a BSc under the same method of work and part time study.[viii]

Leaving a business organisation, she moved to head a professional organisation, that of the Institute of Physics, as Chief Executive, a post she held between 2002 and 2004. She left the professional post for another academic one, as Principal of the Faculty of Engineering at Imperial. She remained in that role until becoming Vice Chancellor of Aston University, Birmingham in 2006.

One area where women are well represented is biomedical engineering Julia asserts. The role of biomedical engineering in solving issues that affect people directly makes it appealing to women, she discusses the course at Imperial College:

> It's a really tough course; it has mechanical engineering, it has electrical and control engineering. It's actually quite similar to aeronautical engineering, except the context isn't an aircraft, it's the human body.
>
> There's also quite a strong environmental interest, and it'd be interesting to see whether environmental engineering, in the sense of green technologies [such as] low-carbon power generation, was starting to attract higher proportions of women.

She admits to feeling "somewhat bitter that medicine has pinched all those women that we might have trained as engineers! With both subjects pulling on the same pool of girls studying science A-levels, it could be that the increase in numbers of medical students has been at the expense of engineering departments."[ix]

Studying materials science would also give students broad options and transferrable skills as its applications are vast. The greatest achievements of the 20th century have all been associated with materials science she says: nuclear energy, flight, discovery of DNA for example.

When Julia was taking her degrees, transferrable skills were not much considered, but today they are very much so. Skills learnt in one area can be applicable to others. Taking a science degree involves many other skills — discipline, integrity in experimentation, methodical investigation, report writing, and there are now many cross subject degrees such as design and engineering, physics and music, besides the biomedical engineering previously mentioned. Such skills also assist with managing workloads, and directing companies. For example, a more administrative than hands on science role can be successfully developed from the basis of a science degree, and Julia herself is evidence of this. She has been involved in a great deal of managerial and advisory work at the highest levels, yet still understands what it is like to be making a career choice and making a first degree choice.

Throughout her career Julia's advice has been sought by authorities in education, engineering (she sometimes wonders if she should have taken engineering as a degree) and government.

She was a member of the government Technology Strategy Board, headed the King Review of vehicle and fuel technologies to reduce carbon emissions over 25 years, which reported in 2008, and is the UK's Low Carbon Business Ambassador.

Julia's career progression has shown how she developed and built on her early ambition of becoming a professor, although she has more than fulfilled that through her work at academic institutions, and by reaching the pinnacle of the academic world, a Vice Chancellor, heading a dynamic and diverse university and being a woman's voice in international science. Her status was further enhanced by her appointment as Baroness Brown of Cambridge, a life peerage in 2015. She took her place in the House of Lords as a crossbencher, and therefore an independent thinker and speaker. Julia says on Aston's website: "It is a huge honour to be appointed to the House of Lords I am looking forward to being able to contribute to debate, and the honing of UK legislation on important areas from my personal experience…" [x] This is a role she will take much pleasure in and give much attention to, as she relinquishes her Vice Chancellorship in September 2016.

Endnotes

[i] *The Guardian*. Retrieved from http://www.theguardian.com/women-in-leadership/women-leadership-blog/2014/oct/20/women-science-engineering-under-representation.

[ii] Julia King, p. 21–23 C1379/43 Track 1 National Life Stories in partnership with the British Library, *An Oral History of British Science Professor Dame Julia King* Interviewed by Dr Thomas Lean on 20110218. Retrieved from http://sounds.bl.uk.

[iii] *Ibid.*, pp. 10–11 C1379/43 Track 1.

[iv] *Ibid.*, p. 44 C1379/43 Track 1.

[v] *Ibid.*, p. 29 C1379/43 Track 1.

[vi] Professor John Knott was awarded the ICF Cottrell Gold Medal 2013 for Outstanding Life Achievement on 1st July 2013 at a special ceremony Aston University, following the ICF13 Conference. Professor Dame Julia King DBE FREng, a former PhD student of John Knott at the University of Cambridge, was awarded the Inaugural ICF Constance Tipper Silver Medal. The *FESI* Bulletin* (*UK Forum for Engineering Structural Integrity), Vol. 7, No. 2, p. 5, Winter 2013.

[vii] The Engineer Women in Engineering Supplement Aston Vice Chancellor Julia King Stuart Nathan 12th July 2015. Retrieved from http://www.theengineer.co.uk/issues/the-engineer-women-in-engineering-supplement/aston-vice-chancellor-julia-king/http://www.theengineer.co.uk/aston-vice-chancellor-julia-king/1010482.article.

[viii] "Why higher apprenticeships? They're a debt-free route to a top quality degree" Julia King, *The Guardian* 14 May 2014.

[ix] *Op cit.*, vii The Engineer Women in Engineering Supplement Aston Vice Chancellor, Julia King Stuart Nathan 12 July 2015. Retrieved from http://www.theengineer.co.uk/aston-vice-chancellor-julia-king/1010482.article.

[x] Vice Chancellor Professor Dame Julia King appointed to the House of Lords. Retrieved from http://www.aston.ac.uk/news/releases/2015/october-2015/vice-chancellor-julia-king-appointed-to-the-house-of-lords/.

Women's Roles: Then and Now

Women's Roles

As College and society evolve, there are necessarily changes in requirements for roles

Many roles have vanished, such as the women 'computers' or those who made the mathematical computations required for research. Beatrice Mabel Cave-Browne-Cave[i] was one such woman, working in the Aeronautics Department in the 1920s. The 'Laboratory Woman' role disappeared many years ago. Now there are highly qualified technicians.

Labs today are filled with very complex machinery and require lab technicians and teaching support staff, many of whom have science degrees. The digital age has changed the way almost everything is done!

This means that a role such as that of Tracer have vanished. The Tracer was a copyist of technical drawings, using opaque paper laid over the original to draw on, and so copy the image onto the opaque sheet, such a role was held by a Miss Mary Wingate in the Civil Engineering Department 1933–1935.

Reprographic machines and computer techniques have mostly taken over from hand-drawn copies, and now there is generally less need for a managed reprographics system, but in 1998 Imperial still had such a system. The administrative department reprographic services were run by Maria Scott, who received the outstanding service award for the services from the Rector. Maria commented "Some of the work I do, particularly with

Maria Scott and the Rector (1993–2000) Lord Ron Oxburgh.

photography, can be complex. It takes a long time to achieve a good result and receiving this award is recognition of that."[ii]

Technical drawings and posters for lecture illustration such as those produced by Margaret Reeks (see entry) have been superseded by PCs and software packages, and the user creates their own illustrative material, and does their own typing!

Rosamund Constance Gunner Sherwood: 'Sherrie' joined imperial in 1919 as a shorthand typist (today, such skills are used by court reporters and note takers). Sherrie became the Rector's Secretary. Later she took upon herself the much needed the role of hostel manageress during WWII, when staff and students remained on College premises as part of the war effort. She became a chaperone when, after WWII, women were allowed 12 places and a common room, in the Beit Hostel, the original hall of residence, and male only, adjacent to the students union building. She generously gave £5 towards the cost of an electric fire for this. It was she who kept an eye on the occupants in the now mixed hostel! She was Imperial's official hostess at the formal dinners.

Today a team of wardens live in the halls and help the students to manage themselves. Woodward Hall is the newest hall of residence, and the first

Woodward Hall Wardening Team

named after an academic woman — Professor Joan Woodward [see separate entry] opened in 2015. The Wardening team at Woodward is part of the College's welfare system: "we are here to help you, to organise social events, and occasionally to enforce discipline within the hall, so that everyone enjoys their time here as much as possible. By day, we are all scientists, working hard within the College, either as members of academic staff or as research students; at night (in our spare time), we become Hall Wardens."[iii]

A completely new way of working is via the College Advanced Hackspace, and it is run by a woman Teaching Fellow, Hackspace Manager, Larissa Kunstel-Tabet who explains: "The Hackspace here at the College allows members from anywhere at Imperial to come and prototype their concepts, either for research, personal projects or entrepreneurial ideas." This includes 3D printing, which is one of the most exciting innovations for small scale work and which has spawned a whole new method of working. It enables small scale trials to be made by those with an entrepreneurial spirit. New workshops continue to be added to the Hackspace and the membership is growing steadily. The group is also aiming to provide a physical space at the White City Campus where people can work on their projects and industry events can be run. Besides this use, links

[iv]Teaching Fellow, Hackspace Manager, Larissa Kunstel-Tabet with a 3D printer.

with the local community at White City can be forged via workshops There's also the potential to bring in the recent alumni who are developing ideas to provide a space where they can bring them forward for about a year. "It will mean they can still feel part of the Imperial family."

Endnotes

[i] Beatrice Mabel Cave-Browne-Cave (1874–1947) Taught mathematics Clapham Girls High School, worked with Karl Pearson UCL, WWI Top Secret government work aeronautics, post WWI Aeronautics Department mathematical calculating work for Leonard Bairstow Zaharoff, Professor of Aviation Imperial College until 1937.

[ii] Imperial College *Reporter* 56, 20th January–2nd February 1998. Retrieved from http://www.imperial.ac.uk/publications/reporterarchive/0056/news04.htm. Accessed 26 July 2016.

[iii] Woodward hall Wardening Team. Retrieved from http://hallsf.imperial.ac.uk/woodward/.

[iv] Larissa Kunstel-Tabet Imperial College Advanced Hackspace by Michael Jones Imperial College Advancement http://www3.imperial.ac.uk/newsandeventspggrp/imperialcollege/newssummary/news_2-3-2016-13-57-15.

Women at the Top

1st Woman Pro-Rector

Mary Ritter

Name and Post Nominals

Mary Ritter
OBE, FHEA (FILT), FRSA, FCGI, FRCPath, FRSB.
First woman Pro-Rector.
First Pro-Rector for Postgraduate Affairs 2004–2005.
First Pro-Rector for Postgraduate and International Affairs 2005–2007.
Pro-Rector International Affairs 2007–2011.

Education

BA (Hons) Zoology, DPhil Immunology University of Oxford; Research Fellowships University of Connecticut, USA; Imperial Cancer Research Fund, London.

Career Roles

Lecturer, Reader, 1982–1991 Professor Immunology Royal Postgraduate Medical School Hammersmith Hospital 1991–2001; Vice Dean education 1992–1997, Assistant Vice Principal Postgraduate Medicine 1997–2000, Director, Graduate School of Life Sciences and Medicine

(GLSM) 2000–2006, Chairman Department of Immunology Faculty of Medicine 2004–2006; Pro-Rector Postgraduate Affairs 2004–2005, Pro-Rector Postgraduate and International Affairs 2005–2007, Pro-Rector International Affairs 2007–2011; Emeritus Professor Immunology Imperial College 2012–; Lee Kong Chian School of Medicine Governing Body 2010–2011; Graduate Academy International Advisory Panel A*STAR (Singapore Agency for Science, Technology and Research) 2007–2011; National Research Foundation Singapore Fellowship Evaluation Panel Member 2015–; RISE High Level Expert Group advising EU Commissioner for Research 2015–2016; CEO 2010–2014, International Ambassador Climate-KIC (Knowledge innovation community) 2014–.

Career Affiliations

1972– Member British Society for Immunology; Fellow Institute for Learning and Teaching in HE became UK Higher Education Academy 2001–; Fellow Royal Society Arts 2004; Fellowship City and Guilds of London Institute 2006; Fellow Royal College Pathologists 2006; Fellow Royal Society Biology; UK Medical Research Council's Non-Clinical Careers Training and Development Panel, the UK Biotechnology and Biological Sciences Research Council's Modular Training for Industry Panel; British Federation of Women Graduates Academic Awards Committee; 2006–2011; Programme Review Committee Cambridge, MIT Institute; PM's Initiative two Committee for HE; Chairman UK Research Councils' "UK GRAD" Steering Committee (Oversees the development and delivery of transferable skills at national level).

Career Award

OBE 2014.

Graduate Schools Creation Imperial College

Founding Director of the Graduate School of Life Sciences and Medicine (GSLSM) in 1999, developing it to its current overarching role providing interdisciplinary research activities, an extensive skills training programme and quality assurance for all postgraduate students in the Faculty of Life Sciences and Faculty of Medicine. The Graduate School of Engineering and Physical Sciences in 2002, was the second such school established and used Mary's pattern. These two graduate schools have now been merged to deliver shared programmes.

Lee Kong Chian School of Medicine

As the Pro-Rector for International Affairs, Mary played a key role in the initial discussions, development and negotiation between Imperial, NTU and Singapore government

Climate, KIC

As her second major project as Pro-Rector of International Affairs, Mary represented Imperial College in the initial discussion, development and negotiation of the application to the European Institute for Innovation and Technology (EIT) for Climate-KIC, and was a member of the winning three-person pitch team in December 2009. She was Imperial's representative on the Climate-KIC Governing Board until her appointment as CEO in December 2010. As CEO, she built Climate-KIC to a partnership of > 200 comprising universities, business and regional/city government. Delivering innovation to address climate change mitigation and adaptation. As International Ambassador, she is working with colleagues in Australia to develop a similar Climate-KIC organisation (to be launched in late 2016) to have global impact, and to develop collaborative and commercial opportunities.

Mary chaired the committee that oversaw the redevelopment of the entire ground floor of Imperial's Library, introducing the quiet and group work areas and the cyber café.

Mary has an interest in academic transferable skills training at Imperial College, and national, European and international level. She chaired the College "Roberts" committee that developed and delivered the transferable skills training programme for all Imperial's PhD students and postdoctoral researchers.

Quote

> Professor Ritter said I am highly honoured and absolutely thrilled.
>
> While the OBE is a personal award, it is also a recognition of all those who have worked with me throughout my scientific research career and made it all possible. The citation is for 'Services to Scientific Research and Innovation' and in the context of innovation, although I have been a serial entrepreneur within the academic context, I am particularly proud of my most recent, and biggest, challenge — leading Climate-KIC.[i]

Immunology

Mary's work in Immunology has been aimed at improving immune recovery following clinical bone marrow transplantation and to the possibility of arresting/ameliorating immune decline in the elderly, and she has published extensively in the area. She has supervised the research of more than 40 PhD and Masters' students, has two patents and several licenced laboratory reagents.

Endnote

[i] New Year's Honours for Imperial researchers and alumni by Sam Wong 2nd January 2014 Imperial College News. Retrieved from http://www3.imperial.ac.uk/newsandeventspggrp/imperialcollege/newssummary/. Accessed 26 June 2016.

Sources: Emeritus Professor Mary Ritter Home page Imperial College Faculty of Medicine, Department of Medicine. Retrieved from http://www.imperial.ac.uk/people/m.ritter.

1st Woman Chairman Imperial College Council

Baroness Eliza (Elizabeth) Lydia Manningham-Buller

Name and Post Nominals

Eliza (Elizabeth) Lydia Manningham-Buller
LG DBC Baroness Manningham-Buller of Northampton.

First woman Chairman, Imperial College Council the Governing Body 2011.
Second woman Deputy Chairman 2009.

Specialism

National Security Services.
Joined Imperial in 2009, Resigned 2015.

Education

Northampton High School, Benenden School, Lady Margaret Hall Oxford University.

Career Roles

Teacher 1971–1974; Security Services Officer MI5 1974–1997; MI5 Deputy Director General 1997–2002; Director General 2002–2007; Wellcome Trust Governor 2008; First Woman Chairman of the Wellcome Trust 2015.

Career Awards

Dame Commander of the order of the Bath 2005; Life Peerage — Baroness Manningham-Buller 2008; Reith Lecturer 2011; Order of the Garter 2014; Co President Chatham House 2015.

Quote

> A woman in a male dominated workplace: I never thought about it. All of the successful women I have met — in finance, law, science, law enforcement, engineering, advertising, the arts — have one characteristic in common: they are busy getting on with the job.[i]

"I have admired Imperial, but since becoming more closely involved with the College, I have been more impressed by its standards of excellence and the contributions it makes to the UK and worldwide. Imperial's staff and students have a wonderful academic and cultural tradition and I look forward to contributing to the College's growth and continued success."[ii]

Endnotes

[i] The Gloss The magazine published with the *Irish Times* 1st Thursday of each month. Article by Aifric Campbell 2nd October 2014.

[ii] Imperial College News Release Wednesday 18th May 2011.

Women at the Top 2007–2016

Comparison of Roles and Numbers of Women Staff in Imperial's Centenary Year 2007 and in 2016

	12/07/2007	12/07/2016	% increase
Academic & Research	798	1,080	35
Clerical	4	Not in use[i]	—
Clinical Academic	56	50	–11
Clinical Research	74	124	68
Learning and Teaching	Not in use[ii]	90	—
Research Nurses	45	81	80
Operational Services	103	190	84
Professional Services	1,372	1,771	29
Technical Services	224	214	–4
Grand Total	**2,676**	**3,600**	**35**

Women have a place in Imperial at all levels in: governance, academic, professional, administration and support services, and student roles.

The total percentage of women staff at Imperial in 2016 is 43.9%.

Women Students

In 1917, there were 32 women students recorded, 10 full-time and 22 part-time. ninety years later, in 2007 the total number of women students was 4,501.

Between 2014/2015 and 2015/2016 there was a 5% increase in women students.

As at July 2016, there are 5,897 women students, including undergraduates and postgraduates:

Faculty of Engineering: 1,785.
Faculty of Medicine: 1,616.
Faculty of Natural Sciences: 1,799.
Total Student number: 15,734.

College Structure

Beginning with governance; the structure is The Council, The Court, The Senate[iii]; women have been members of all bodies for many years. The Council is the governing and executive body. The Honorable Sara Morrison was member of the Council 1986–2002, and the first woman Deputy Chairman of Governors. The second woman Deputy Chairman was Eileen Buttle, 2002–2009, also Acting Chairman 2004–2005. Eliza Manningham-Buller became the first woman Chairman in 2011, having been Deputy Chairman since 2009.

The governing body appoints the President as Chief Executive to lead the College.

We now have as Imperial's leader, a woman President, Professor Alice Gast, appointed in 2014.

Beneath these roles are others held by women who lead Imperial as part of governance, as Associate Provosts and Vice Presidents, then as heads and lead roles within academic departments, and also as heads of the professional and support services. The Provost, the second highest role at Imperial is held by Professor James Stirling. The student president is elected annually. The 2015–2016, President of the Student's Union, was a woman, Lucinda Sandon-Allum, the 11th woman President since 1907 (see separate entry).

The presence of women at every level within Imperial College is now obvious, and vital to the achievement of the desired strengths of collegiality and mission.

Associate Provost and Vice President Roles

Professor Maggie Dallman

BSc, DPhil, FCGI FRSB (FSB), OBE.
Associate Provost (Academic Partnerships).
First Associate Provost (Academic Partnerships) 2014–
Joined Imperial 1994.

Education

BSc Hons Cellular Pathology Bristol University 1978; DPhil 1982; MA Linacre College Oxford University FRSB (FSB 2011) 1985.

Career Roles

Research Scientist; 1978–1979 Wellcome Research Labs; Post-doctoral Research Fellow Stanford University 1982–1984; Nuffield Foundation Junior Research Fellow Hertford College 1984–1987; Oxford (Emeritus Fellow 1994); Lecturer Biology Department 1994; Reader in Immunoregulation 1996; Professor of Immunology 1999; Head Section Immunology and Infection 2001; Campus Dean Deputy Principal for the Faculty of Natural Sciences, 2008; Dean, 2010. Instituted a Faculty of Natural Sciences Award for Excellence, Associate Provost (Academic Partnerships) leading Outreach Strategy and International relations 2014–. Chair Imperial's Animal Welfare and Ethical Review Body Imperial College.

Affiliations

Director and Trustee of the Francis Crick Institute, Member, Council BBSRC; Member, the Scientific Advisory Board Pirbright Institute; Member, Board NC3Rs; member HEFC Research and Innovation Strategy

Advisory Committee; Deputy Chair of the Healing Foundation Research Council.

Women's Support

2011 Gave the Imperial College Athena Lecture.

Hosted *Women's Leadership in Entrepreneurial and STEMM Culture 2014*, the event brought together high achieving women from across the higher education, business and the public sectors to share their personal career journeys.

Besides focussing on her Associate Provosts role and advocacy for women, Maggie runs a very active research lab: centring on immune responses to pathogens, studying zebrafish to humans, to investigate at the molecular, cellular and whole individual levels, the basis for disease and potential approaches to therapy.

Quote

> On award of the OBE for services to bioscience:
>
> The award reflects, I believe, my research and educational contributions in the biosciences, together with my work with agencies including the BBSRC, NC3Rs and medical charities — but none of this would have been possible without the many wonderful colleagues I have been inspired by and had the privilege of working with over the years both before and since moving to Imperial.

Sarah Porter Waterbury Vice President Advancement

First Vice President (Advancement) 2015–.

Role

Responsible for the College's Advancement function, encompassing all aspects of alumni relations, events and development. Member of the President's Executive Group.[iv]

Maggie Dallman (centre) with her ZebraFish Lab team.

Education

BA Princeton University.

Career Roles

JP Morgan Private Banking Division; Major Gifts Officer, Harvard Business School 1993, Director, Development and External Relations Fletcher School Tufts University; Deputy Vice President Development American University 1999; Senior Director of Campaign and Major Gifts at Children's Hospital Boston 2000; Vice President Development and Campaigns New York University 2009.

Quote

> People are people, the world over… and people like to give to people… They need to believe that we will act responsibly and use their gift

appropriately and in accordance with their wishes. That's all about trust…Ultimately we're trying to showcase the university to the right audiences, whether it's people or organisations that can help us with our financial and resource needs or recommend us to prospective staff and students and tell them what a great place it is to work and study.[v]

Endnotes

[i] Not in Use 2016: Reflects how College responds to changing requirements in the organisation and external trends and requirements.

[ii] Not in use 2007 *ibid.*, i

[iii] College Governance Structure. Retrieved from: http://www.imperial.ac.uk/admin-services/secretariat/college-governance/governance-structure/ Accessed 23 July 2016.

[iv] Imperial College Leadership and Strategy. Retrieved from: https://www.imperial.ac.uk/about/leadership-and-strategy/president/vice-president-advancement/

[v] Profile: Sarah Porter Waterbury, Imperial's first Vice President of Advancement by Andrew Czyzewski 1st May 2015. Retrieved from: http://www3.imperial.ac.uk/newsandeventspggrp/imperialcollege/newssummary/news_1-5-2015-10-25-13.

Biography

Alice Petry Gast

Name and Post Nominals

Alice Petry Gast (1958–)

PhD.

President, Imperial College London.

First Woman President Imperial College.

First Woman President Lehigh University.

First woman to create Women's STEM Scholarship Fund at Lehigh.

Specialism

Chemical Engineering, colloid science, polymer physics, statistical mechanics, microfluidics; University administration and leadership.

Career Roles

Chemical Engineer, Lecturer, Professor Stanford Synchrotron Radiation Laboratory. 1985–2001 Stanford University; Associate Provost and Vice President for Research Robert T. Haslam Chair in Chemical Engineering Massachusetts Institute of Technology 2002–2005; President, Lehigh

University 2006–2014[i]; National Research Council Board on Chemical Science and Technology; Homeland Security Science and Technology Advisory Committee; Chaired the Committee on *Science and Security in a Post 911 World: A Report Based on Regional Discussions Between the Science and Security Communities Committee on a New Government* (University Partnership for Science and Security) 2006 National Academy of Sciences; International Science Envoy 2010 USA Government; Chaired the Committee on Review of the Scientific Approaches Used During the FBI's Investigation of the 2001 *Bacillus Anthracis* Mailings 2011.[ii]

Career Awards

Award for Initiative in Research National Academy of Sciences 1992; Colburn Award of the American Institute of Chemical Engineers 1992; Dreyfus Teacher Scholar Award[iii]; Guggenheim Fellowship 1991[iv] Alexander von Humboldt Fellowship, held at Technisch Univerisität Munchen 1999; Honorary Degree, University of Western Ontario 2010; Honorary Chair East China University of Science and Technology (ECUST) 2012; Alice P. Gast Women in STEM Scholarship Lehigh University 2012; Ambassador Technisch Univerisität Munchen 2015.

Awards

One of top 100 Modern Era engineers, Leadership category, American Institute of Chemical Engineers.

Professional affiliations

Member, Board, American Association for the Advancement of Science (AAAS); the National Research Council Committee for Science, Technology, and the Law National Academy of Engineering 2001; American Academy of Arts and Sciences 2002; member of the Academic Research Council for the Singapore Ministry of Education; American Chemical Society; American Institute of Chemical Engineers; American

Physical Society; U.S. Manufacturing Competitiveness Initiative Steering Committee Council on Competitiveness; Member, Audit Committee Board Chevron Corporation 2012; Member, Board of Trustees for the King Abdullah University of Science and Technology (KAUST) Thuwal Saudi Arabia; Member, Academic Research Council for the Singapore Ministry of Education; Member, Global Science and Innovation Advisory Council to the Prime Minister of Malaysia; Member, Leadership Council National Centre for Universities and Business (NCUB) UK.

Lectures

The Global Economic Crisis and Higher Education at the Lord Dearing Memorial Conference 2010,[v] Athena Lecture Order, Disorder, Complex Fluids & Life, Imperial College 2016.

Quote

> "At Imperial students and staff are on a common mission, they interact very well together and are very collaborative."[vi]
>
> "Always try to simplify questions, mix different viewpoints, collaborate near and far, and value your mentors, students, friends, colleagues and family"[vii]

Born

25th May 1958 Houston Texas USA.

Family Background

Father: Biochemist.

Marriage

Bradley J Askins Computer Scientist, Children one son David, one daughter Rebecca

Education

University

Undergraduate

University of Southern California.

Degree

1980 BSc Valedictorian Chemical Engineering.

Postgraduate

Princeton University.

Degree

MA 1981, PhD 1984.

Postgraduate Awards/scholarships

Hertz Fellowship.

Postdoctoral

NATO fellowship at the École Supérieure de Physique et de Chimie Industrielles in Paris.

Text Discussing Career and Other Comments

Alice was born into a scientific family, her father was a biochemist and she remembers being in the labs and the excitement about an electron microscope revealing cell structures. All this was absorbed for use in her later career. She was the child who changed the spark plugs on the family car, and was very much an athlete, excelling in the long jump. Science, engineering and athletic pursuits have informed Alice's life, along with close family bonds.

Throughout her career, Alice has led from the front and has developed successful research groups such as the Gast Group at Stanford University and from hands on science has gone on to develop a societal view of what science can do for the world and humanity. This has been recognised by her many invitations to join US national and international committees of great import for future developments in the scientific, higher education and security debates in the US and elsewhere. The most profound acknowledgment of her valued status is that she was selected as one of only three new International Science Envoys in 2010 by the Secretary of State, Hilary Rodham Clinton. This led to extensive travel in Central Asia and the Caucasus, and to Alice advising the US Government on strengthening relations with those countries. That she has been awarded an honorary professorship by East China University of Science and Technology is testament to the impact of her worldwide science advocacy. Similarly, her membership of Board of Trustees for the King Abdullah University of Science and Technology Saudi Arabia, the Academic Research Council for the Singapore Ministry of Education and the Global Science and Innovation Advisory Council to the Prime Minister of Malaysia also demonstrates that her diplomacy as a women scientist has paid dividends.

At Lehigh University in Bethlehem Pennsylvania, Alice led much expansion during her tenure as President 2007–2014, in developing land as new campus sites, interdisciplinary research facilities and in new academic opportunities worldwide, and very successfully increased the endowment fund. She introduced STEPS, an initiative on science, technology, environment, policy and society, and oversaw Lehigh's Environmental Advisory Group and emergency and crisis management planning, including for environmental disasters. She also took a stand on high risk behaviours amongst students, such as alcohol consumption and was a member of the Council of the National Institute on Alcohol Abuse and Alcoholism (NIAAA) in 2010.[viii]

Alice understands how important it is to support women entering, and indeed, remaining in science, technology, engineering and maths, and to further this cause, established a million dollar research fund Lehigh in 2012, which was named the Alice P. Gast Women in STEM Scholarship as a legacy title on her leaving Lehigh for Imperial.

Her experience at Lehigh and other roles has stood her in good stead for leading Imperial through the challenges of an international academic institution in the 21st century.

Giving the vote of thanks after Alice's Athena Lecture at Imperial, the Provost James Stirling said

> The label 'chemical engineer' fails to do justice to the breadth of Alice's expertise. When we consider the range of objects she has studied and the techniques her research has utilised, we see a chemist, a physicist, a biologist, a mathematician, a materials scientist and an engineer at work. She epitomises what Imperial is all about.[ix]

However, as her Athena Lecture also attested as she described her life, Alice is a very down to earth person, her hobby is baking which helps her wind down, as she relaxes with her family which includes three cats. The human members of her family all enjoy exploring London and holidaying together, when their schedules, all very busy and on at least two, possibly three continents, permit. Alice is approachable and friendly and keen to make Imperial her academic family, encompassing all who study and work here. Alice fosters the collaborative spirit which will continue Imperial's trajectory as a leading global university.

Endnotes

[i] Lehigh University Bethlehem Pennsylvania.

[ii] Chaired the scientific review committee National Academy of Sciences Research Council 2011 into the Amerithrax case. At the request of the FBI independent review of FBI investigatory methods in the 2001 criminal case of anthrax spores being posted to individuals. Retrieved from http://www.nap.ed http://www.nap.edu/catalog/13098/review-of-the-scientific-approaches-used-during-the-fbis-investigation-of-the-2001-anthrax-letters.

[iii] Camille Dreyfus Teacher-Scholar Awards Program for early career chemical scientists showing leadership skills in scholarly research nominated by their institution. The Camille and Henry Dreyfus Foundation Inc. Retrieved from http://www.dreyfus.org/awards/camille_dreyfus_teacher_award.shtml Accessed 20160718.

[iv] John Simon Guggenheim Memorial Foundation grants awarded annually to those "who have demonstrated exceptional capacity for productive scholarship or exceptional creative ability in the arts." http://www.gf.org/about/fellowship/.

[v] The Lord Dearing Memorial Conference *'The Future of Higher Education' 2010*. To mark the life and achievements in Higher Education strategy of Lord Ron Dearing (1930–2009) Chancellor Nottingham University 1993–2000. Chairman of the committee on *Higher Education in the learning Society* (The Dearing report) 1997. Retrieved from http://www.alumni.nottingham.ac.uk/NetCommunity/Page.aspx?pid=1436. Accessed 20160718.

[vi] Interview with Imperial College Students Union President 2015.

[vii] Alice Gast *Order, Disorder, Complex Fluids & Life* Athena Lecture Imperial College, 1st June 2016 Imperial College *Reporter* 2nd June 2016 Deborah Evanson.

[viii] Lehigh University. Retrieved from http://www1.lehigh.edu/about/history/gast Accessed 20160718.

[ix] *Ibid.*, vii.

Afterword

Imperial College London's mission is to achieve enduring excellence in research and education in science, engineering, medicine and business for the benefit of society. As students, support staff and academics, women have significantly added to Imperial's culture of excellence and to its mission. These contributions are testament to their brilliance and hard work and to the forward thinking of this great institution. Imperial embraces diversity.

The women featured in this book are inspirational. From the early days of Imperial to the present, women have been leaders and pathfinders, each generation laying a foundation for the next. Their biographies remind us that great discoveries begin with great people: talented individuals steeped in the knowledge of their core discipline, confident enough to work on risky, unsolved problems, adept at understanding and working with others from different fields and immersed in an atmosphere of excellence.

In addition to the women included in this history, there are many, many others whose service and dedication to Imperial helped make it the leading university that it is today.

- Our talented support staff put their full energy and passion into their work and our mission. Their impact is enormous and felt every day at Imperial.
- Our bright and energetic students are the leaders of the future. They will become the next generation of academics, entrepreneurs and leaders.

- Our alumni and friends are a tremendous asset. Their knowledge, expertise and experience contribute to the intellectual power of the college and their financial support helps us in our pursuit of break-through solutions to complex societal problems.
- We welcome people from across the globe to Imperial College and we work with schools and invite the public to learn more about science, engineering, medicine and business, Imperial Fringe Fairs and Festivals. Our website will always have details: https://www.imperial.ac.uk and a cordial and expansive scientific experience awaits you.

I see women playing an even larger role at Imperial in the years to come. We will continue to provide the facilities and professional support services for the development of women in science, engineering, medicine and business. We will continue to do all we can to attract women students and staff and create an environment where they can thrive. We will continue the proud tradition of Women at Imperial.

The history of women at Imperial is impressive, and I am honoured to be part of that history and am privileged to be the first woman President. I am certain that I will not be the last.

Professor Alice Gast
President Imperial College, London

Exhibition panel for Women@Imperial Week featuring Emily Chapman when a PhD student 2012.

Appendix I

Abbreviations Used

AIP	American Institute of Physics
APS	American Physical Society
ARSM	Associateship Royal School of Mines
ARCS	Associateship Royal College of Science
AC&GC	Associateship City and Guilds College
ASGBI	Association of Surgeons of Great Britain and Ireland
AWISE	Association of Women in Science and Engineering
BAAS	British Association for the Advancement of Science now British Science Association
BSc	Bachelor of Science, undergraduate degree
BMA	British Medical Association
BM BCh	Bachelor of Medicine, Bachelor of Surgery (Oxford University)
BSA	British Science Association formerly British Association for the Advancement of Science
BTA	'Been to America' medical training
CBE	Commander British Empire

CChem	Chartered Chemist, Royal Society of Chemistry professional qualification
CH	The Order of the Companions of Honour (Reward for outstanding achievements in the arts, literature, music, science, politics, industry and religion)
Ch.M	Master of Surgery (Magister Chirurgiae in Latin)
Chevalier of the Légion d'honneur	Knight of the National Order of the Legion of Honour (French award)
CO-PI	Joint Principal Investigator
DBE	Dame British Empire — Dame Commander of the Most Excellent Order of the British Empire, a grade within the British order of chivalry
DIC	Diploma Imperial College
DPhil	Doctor of Philosophy
DSIR	Department of Scientific and industrial Research
EUROMAT	European Collaboration in Measurement Standards
EURAMET	European Association of National Metrology Institutes formerly EUROMAT
ESDU	Engineering Science Data Unit now IHS ESDU source of authoritative information for engineering design
ESPRC	Engineering and physical Science Research Council
FACC	Fellowship of the American College of Cardiology
FAHA	Fellowship of the American Heart Association
FAPS	Fellowship of the American Physical Society
FASCS	Faculty of Arts and Sciences Computer Services
FDSRCS	Fellowship of the Dental Surgery of the Royal College of Surgeons

FIM	Fellowship of the Institute of Metals (Now FIMM)
FIMarEST	Fellowship of the Institute of Marine Engineering Science and Technology
FIMMM	Fellowship of the Institute of Materials Minerals and Mining (IOM3) (Merger of Institute of Materials and Institution of Mining and Metallurgy)
FInstP	Fellowship of the Institute of Physics
FINucE	Fellowship of the Institution of Nuclear Engineers
FMEdSci	Fellowship of the Academy of Medical Sciences
FRAeS	Fellowship of the Royal Aeronautical Society
FRCP	Fellowship of the Royal College of Physicians
FECISHR	Fellowship of the European Council of the International Society for Heart Research
FESC	Fellowship of the European Society of Cardiology
FPS	Fellowship of the Pharmacological Society
FREng	Fellowship of the Royal Academy of Engineering
FRS	Fellowship of the Royal Society (of London for Improving Natural Knowledge)
FRSC	Fellowship of the Royal Society of Chemistry
HSE	Health and Safety Executive
ICE	Institution of Civil Engineers
ICF	International Conference on Fracture
IDHDP	International Doctors for Healthier Drugs Policies
ICSU	International Council of Scientific Unions

IDHDP	International Doctors for Healthier Drugs Policies
IEE	Institution of Electrical Engineers now IeT Institution of Engineering Technologies
IEEE	Institute of Electrical and Electronic Engineers (USA)
IeT	Institution of Engineering Technologies formerly IEE Institution of Electrical Engineers
IHS ESDU	see ESDU
INucE	Institute of Nuclear Engineers
IoP	Institute of Physics
IUPAC	International Union of Pure and Applied Chemistry
LRCP	Diploma of the Licentiateship of the Royal College of Physicians no longer used
MRCS LRCP	conjoint qualifications of the Royal College of Physicians and Royal College of surgeons between 1885 and 2001 no longer used
MA	Master of Arts
MBCHB	Bachelor of Medicine, Bachelor of Surgery Latin: Medicinae Baccalaureus, Baccalaureus Chirurgiae
MBBS	Bachelor of Medicine Bachelor of Surgery
MD	Medical Doctor
MRC	Medical Research Council
MRCP	Member Royal College of Physicians
MRCS LRCP	conjoint qualifications of the Royal College of Physicians and Royal College of surgeons between 1885 and 2001 no longer used
MRCS	Member Royal College of Surgeons
NIRAB	Nuclear Innovation and Research Advisory Board

OBE	Order of the British Empire (Most Excellent Order of the British Empire in the Order of Chivalry)
OM	Order of Merit
PG	Postgraduate
PGR	Postgraduate Research
PPARC	Particle Physics and Astronomy Research Council
RSC(E)	Royal College of Surgeons of England
SRC	Science Research Council
IFC	International Fatigue Congress
STEM	Science, Technology Engineering and Maths
UG	Undergraduate
UK	United Kingdom of Great Britain and Northern Ireland
UKRC	UK Resource Centre for Women now incorporated into WISE. The UKRC is a UK organisation for the provision of advice, services and policy consultation regarding the under-representation of women in science, engineering, technology and the built environment (SET) (Available at: http://www.setwomenstats.org.uk/)
UKRC	Radiological Science UK annual multidisciplinary scientific congress and technical exhibition organised by the British Institute of Radiology, the Society and College of Radiographer and the Institute of Physics and Engineering in Medicine
UKRMP	UK Regenerative Medicine Platform
VSS	Vascular Society of Great Britain and Ireland

WISE	WISE Campaign Women into Science and Engineering now incorporated into UKRC https://www.wisecampaign.org.uk/about-us
WiST	Women in Surgical Training Scheme

Explanation of other titles:

Lambeth Degrees (Cicely Saunders). Available in http://www.academia.edu/1991173/Lambeth_Degree_Academical_Dress

Appendix II

Degrees for Women

University of London

Granted a Supplemental Charter in 1878 to admit women as candidates for degrees on equal terms with men. Imperial College Associateship awards gained UoL internal degree status in 1926. Imperial College gained degree giving powers on independence from UoL in 2007.

Cambridge University

1868: Women's Local Examination, later the Higher Local Examination, instituted by the Senate of Cambridge University.

1869: Henry Sidgwick, Fellow of Trinity College, began a Committee to devise courses for the examinations, which in 1873 became the Association for the Promotion of the Higher Education of Women in Cambridge.

1881: Women students were admitted to the Tripos (and Previous) examinations, receiving a certificate in place of a degree.

1923: Women at Cambridge received titular degrees.

1948: Full membership of the University.[i]

Oxford University

1870s: Women could not matriculate (be recognised as members of the university) but could take course and sit exams. They were not awarded a degree for their examinations, although they were marked and it was clear that they had achieved honours status.

1920: University statute admitted women to full membership of the University, and retrospectively gave matriculation and degrees to those eligible. Imperial College Associateship awards gained.[ii]

Endnotes

[i] British History Online. Retrieved from http://www.british-history.ac.uk/vch/cambs/vol3/pp493-495.

[ii] University of Oxford Archives. Retrieved from http://www.oua.ox.ac.uk/enquiries/first%20woman%20graduate.html.

Illustrations

Introduction

Linda Jones	Imperial College
Mariam Oommen	Imperial College Archives
Judith Walker	Judith Walker
Royal School of Mines Museum of Practical Geology viewed by ladies accompanied by Professor Edward Forbes and Curator/Registrar Trenham Reeks 1851	Imperial College Archives
T.H. Huxley and wife inland sea dredging on their honeymoon	Imperial College Archives

Biographies

Lady Rachel MacRobert Workman	MacRobert Estate
Professor Janet Vida Watson	Imperial College
Professor Dame Julia Stretton Higgins	Imperial College
Professor Anita Bailey	Imperial College
Photograph Chemical Engineering Department. 1984 Professors 5 years apart: Dame Julia Stretton Higgins and Professor Anita Bailey. Marjorie De Reuck IUPAC unit	Imperial College Archives

Letitia Chitty	Cambridge University
Constance Fligg Elam Tipper	Imperial College Archives
Assistant Professor Martha Annie Whiteley	Imperial College Archives
Letter concerning 'Common Cause' scholarship	Imperial College Archives
Lecture Notebook Cover 1897	Imperial College Archives
Martha Annie Whiteley Retirement Book Testimonial	Imperial College Archives
Nellie Ivy Fisher and Frances Mary Hamer	Imperial College Archives North London Collegiate School
Ida Smedley Maclean	The Wellcome Library CC
Professor Helen Kemp Porter	Imperial College Archives

Margaret Lacey and Biology Department colleagues. Folk dancing on Lab block Roof Beit Quad, 1920s.

Elsie May Widdowson	Imperial College
Professor Dame Julia Polak	Imperial College
Dame Janet Vaughan	National Portrait Gallery Bassano
Winifred May Watkins	Biochemical Society
Professor Celia Oakley	Courtesy her daughter Sue Soar and Royal Free Association
Professor Dame Margaret Turner Warwick	Imperial College By her Jeff Stultienes portrait
Professor Dame Josephine Barnes	Wellcome Library London CC

Professor Lesley Regan	Imperial College
Professor Averil Mansfield	Imperial College
Dame Cicely Saunders	Grace Goldin King's College London
Edith Summerskill	Imperial College
Jane Wadsworth	Imperial College

Meteorological Office South Kensington 1925. Elen Elaine Austin signed as EEA in manuscript text, far right. Copyright owner undiscovered

Daphne Jackson	New Scientist
Hertha Ayrton First Woman Royal Society Hughes Medal Winner	Imperial College Archives
Michele Dougherty Second Woman Royal Society Hughes Medal Winner	Imperial College
Joanna Haigh	Imperial College
Dame Sue Ion	Imperial College
Reeks Family Trenham Reeks	British Geological Survey
Margaret Reeks Notebook Illustrations	Archives Imperial College
Judd Family John Wesley Judd, Hilda Judd	Imperial College Archives
Conrady family Eugen, Hilda, Doris, Conrady and Rudolf Kingslake	Imperial College Archives
Hill Family Peggy, Ted and Allen Hill	Imperial College Archives
Chain Family Anne Beloff Lady Chain and Sir Ernst Chain	Imperial College
Sergot Family Teresa and Marek: Antoni and Leon Sergot	Sergot family

Hankin Family: Chris Hankin and Alison Holtorp:

Emily Martin née Hankin, Richard Martin, David Hankin and Anna Hankin née Kovaleva	Hankin Family

Liberal Studies and History of Science

Touchstone weekend Silwood Park	Imperial College Archives
Magda Whitrow	Robert Williams Carolina University

Pastoral Roles

Sue Thornett	Imperial College Archives
Margaret Cunningham and Lynda White	Imperial College
Anna Thomas-Betts	Imperial College
Margaret Goodgame	Imperial College
Lorraine Craig	Imperial College
Judy Barnett	Imperial College
Sian Harding	Imperial College (Jackie King)

Women's Clubs — Social

President Imperial College Women's Club, Lady Vincent, and Professor Julia Higgins	Imperial College

Imperial College Women's Association (ICWA)

Original Imperial Women's Association	Imperial College Archives
1924 President from Phoenix	Imperial College Archives
1925 ICWA Tea Party	Imperial College Archives
ICWA Then and Now 1974	Imperial College Archives

Tie Clubs and Women

Toni Semmence Founder & First President the XV	Toni Semmence
The XV Tie bracelet	Toni Semmence

Women Students Union Presidents and Roles

2015–2016 President's Address Lucinda Sandon-Allum	Imperial College

Recruiting Women Students

Recruitment brochures	Imperial College Archives

Centenary Prize Lists

Yael Naim Dowker	MFO Oberwolfach Photo Collection CC
Brenda Ryman	Girton College Cambridge

Women@Imperial: Gender Equality, Athena Swan and Women Ambassadors, Scholarships, Scientific and Entrepreneurial Initiatives at Imperial and for Returners to Scientific Careers, Women@Imperial Week

Sue Gibson	Imperial College
Dr Liz Elvidge	Imperial College
Maria Petrou	Imperial College
Maria Belvisi	Imperial College
Lesley Cohen	Imperial College
Ann Muggeridge	Imperial College
Karen Makuch	Imperial College
Victoria Salem	Imperial College
Diane Morgan	Imperial College
Dorothy Griffiths	Imperial College
Anne Dell	Imperial College

Industrial Sociology to Business School

Professor Dorothy Wedderburn	Imperial College

Professor Joan Woodward	Imperial College
Schoolgirl league of Nations Essay Prize Winner	Oxford Times
Professor Dame Sandra Dawson	Cambridge University
Professor Sue Birley	Imperial College
Professor Dorothy Griffiths OBE	Imperial College

Principals and Vice Chancellors

Frances Micklethwait Principal Swanley Horticultural College	Imperial College Archives
Kate Barratt Principal Swanley Horticultural College	Imperial College Archives
Professor Jenny Higham Principal St Georges Medical School London	Imperial College
Professor Debra Humphris VC Brighton University	Imperial College
Professor Julia Clare Buckingham VC Brunel University	Imperial College
Baroness Brown, Julia King (Formerly) VC Aston University	Imperial College

Women At the Top

Mary Ritter	Imperial College
Baroness Manningham-Buller	Imperial College
Maggie Dallman	Imperial College
Sarah Porter Waterbury	Imperial College
Biography Alice Gast	Imperial College
Image of Women@Imperial week	Imperial College

Index

B

H

L

O

T

U